I0054788

Raffaele Chiappinelli

Differential Equations, Fourier Series, and Hilbert Spaces

Also of Interest

Fourier Meets Hilbert and Riesz
An Introduction to the Corresponding Transforms
René Erlin Castillo, 2022
ISBN 978-3-11-078405-3, e-ISBN (PDF) 978-3-11-078409-1
in: De Gruyter Studies in Mathematics
ISSN 0179-0986

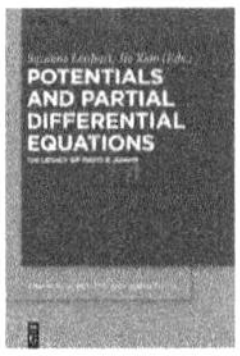

Potentials and Partial Differential Equations
The Legacy of David R. Adams
Edited by Suzanne Lenhart, Jie Xiao, 2023
ISBN 978-3-11-079265-2, e-ISBN (PDF) 978-3-11-079272-0
in: Advances in Analysis and Geometry
ISSN 2511-0438

Topics in Complex Analysis
Dan Romik, 2023
ISBN 978-3-11-079678-0, e-ISBN (PDF) 978-3-11-079681-0

Advanced Mathematics
An Invitation in Preparation for Graduate School
Patrick Guidotti, 2022
ISBN 978-3-11-078085-7, e-ISBN (PDF) 978-3-11-078092-5

Differential Equations
A first course on ODE and a brief introduction to PDE
Shair Ahmad, Antonio Ambrosetti, 2023
ISBN 978-3-11-118524-8, e-ISBN (PDF) 978-3-11-118567-5

Raffaele Chiappinelli

Differential Equations, Fourier Series, and Hilbert Spaces

Lecture Notes at the University of Siena

DE GRUYTER

Mathematics Subject Classification 2020
Primary: 41-01, 46-01; Secondary: 34-01, 35-01, 54-01

Author
Raffaele Chiappinelli
University of Siena
Department of Information Engineering and
Mathematical Sciences (DIISM)
Via Roma 56
53100 Siena
Italy
raffaele.chiappinelli@unisi.it

Figures by Rita Nugari

ISBN 978-3-11-129485-8
e-ISBN (PDF) 978-3-11-130252-2
e-ISBN (EPUB) 978-3-11-130286-7

Library of Congress Control Number: 2023940856

Bibliographic information published by the Deutsche Nationalbibliothek
The Deutsche Nationalbibliothek lists this publication in the Deutsche Nationalbibliografie; detailed bibliographic data are available on the Internet at http://dnb.dnb.de.

Cover image: RED_SPY / iStock / Getty Images Plus
Typesetting: VTeX UAB, Lithuania
Printing and binding: CPI books GmbH, Leck

www.degruyter.com

Introduction

This book is intended to be used as a rather informal, and surely not complete, textbook on the subjects indicated in the title. It collects my lecture notes held during three academic years at the University of Siena for a one-semester course on “basic mathematical physics,” and is organized as a short presentation of a few important points on the arguments indicated in the title, the interested audience being advanced (third year) undergraduate or first year postgraduate students in mathematics, physics, or engineering.

Chapter 1 and Chapter 4 of this book are devoted to, respectively, completing the students’ basic knowledge on ordinary differential equations (ODEs), dealing in particular with those of higher order, and providing an elementary presentation of the partial differential equations (PDEs) of mathematical physics, by means of the classical methods of separation of variables and Fourier series. For a reasonable and consistent discussion of the latter argument, some elementary results on Hilbert spaces and series expansion in orthonormal vectors are treated with some detail in Chapter 3.

In brief, my hope is that the present notes can serve as a second quick reading on the theme of ODEs and as a first introductory reading on Fourier series and Hilbert spaces and on PDEs. A complete discussion of the results on ODEs and PDEs that are here just sketched is to be found in other books, specifically and more deeply devoted to these subjects, some of which are listed in the Bibliography (see for instance [1] and [2], respectively; see also the recent [3], covering both subjects).

Prerequisites for a satisfactory reading of the present notes are not only a course of calculus for functions of one or several variables and linear algebra (as treated, for instance, in [4]), but also a course in mathematical analysis where – among others – some basic knowledge of the topology of normed spaces is supposed to be included. At any rate, during the final preparation of the manuscript I decided to insert a further chapter dealing with metric and normed spaces, adding it to the originally planned three chapters of the book whose content I have indicated before. The goal of this addition is also that of preparing the possibly interested student (typically, a mathematics undergraduate) to follow in his subsequent career a course in functional analysis; some familiarity with metric and normed spaces is clearly a *conditio sine qua non* to appreciate the content of such kind of course. In any case, the study of this additional chapter (appearing in the book as Chapter 2) is in no way strictly necessary for the understanding of the remaining three chapters; I have paid particular attention to organize things so that this study be confined to the role of additional sources of information for those interested in a deeper analysis of the statements and results shown in Chapters 1, 3, and 4.

Siena, May 2023 Raffaele Chiappinelli

https://doi.org/10.1515/9783111302522-201

Contents

Preliminaries

Topology in $\mathbb{R}^n$: balls, interior points, open sets

For each vector $\boldsymbol{x} = (x_1, \ldots, x_n) \in \mathbb{R}^n$ put

$$\|\boldsymbol{x}\| = \sqrt{\sum_{i=1}^{n} x_i^2} \tag{0.0.1}$$

and call $\|\boldsymbol{x}\|$ the (Euclidean) *norm* of $\boldsymbol{x}$. Recall that:
- $\|\boldsymbol{x}\| \geq 0$ for all $\boldsymbol{x} \in \mathbb{R}^n$, and $\|\boldsymbol{x}\| = 0$ iff $\boldsymbol{x} = 0$;
- $\|\lambda\boldsymbol{x}\| = |\lambda|\|\boldsymbol{x}\|$ for all $\lambda \in \mathbb{R}$ and all $\boldsymbol{x} \in \mathbb{R}^n$;
- $\|\boldsymbol{x} + \boldsymbol{y}\| \leq \|\boldsymbol{x}\| + \|\boldsymbol{y}\|$ for all $\boldsymbol{x}, \boldsymbol{y} \in \mathbb{R}^n$.

Of the above three properties of the norm, the last is called the *triangle property* and is the only one non-trivial to check on the sole basis of (0.0.1). In fact, it is a consequence of the *Cauchy–Schwarz inequality*:

$$|\boldsymbol{x} \cdot \boldsymbol{y}| \leq \|\boldsymbol{x}\|\|\boldsymbol{y}\|, \tag{0.0.2}$$

where $\boldsymbol{x} \cdot \boldsymbol{y}$ is the *scalar product* of $\boldsymbol{x}$ and $\boldsymbol{y}$, defined putting

$$\boldsymbol{x} \cdot \boldsymbol{y} = \sum_{i=1}^{n} x_i y_i$$

for $\boldsymbol{x} = (x_1, \ldots, x_n)$ and $\boldsymbol{y} = (y_1, \ldots, y_n)$; these statements will be proved in the more general context of inner product spaces (Chapter 3, Section 3.1).

We see from (0.0.1) that

$$|x_i| \leq \|\boldsymbol{x}\| \quad \forall i = 1, \ldots, n. \tag{0.0.3}$$

Given $\boldsymbol{x}_0 \in \mathbb{R}^n$ and $r > 0$, the set

$$B(\boldsymbol{x}_0, r) = \{\boldsymbol{x} \in \mathbb{R}^n : \|\boldsymbol{x} - \boldsymbol{x}_0\| < r\}$$

is called the *spherical neighborhood* (or simply the *ball*) of center $\boldsymbol{x}_0$ and radius r. Thus, when $n = 1$, $B(x_0, r)$ is merely the interval $]x_0 - r, x_0 + r[$. A point $\boldsymbol{x} \in \mathbb{R}^n$ is said to be *interior* to a subset A of $\mathbb{R}^n$ if there exists an $r > 0$ such that $B(\boldsymbol{x}, r) \subset A$. A is said to be *open* if any point of A is interior to A.

Example 0.0.1. The set

$$A = \{\boldsymbol{x} = (x, y) \in \mathbb{R}^2 : x > 0, y > 0\}$$

https://doi.org/10.1515/9783111302522-202

is open; indeed, given any $\mathbf{x}_0 = (x_0, y_0) \in A$, letting $r_0 = \min\{x_0, y_0\}$, we check that $B(\mathbf{x}_0, r_0) \subset A$. To see this, note that if $\mathbf{x} = (x, y) \in B(\mathbf{x}_0, r_0)$, then by (0.0.3) we have

$$|x - x_0| \leq \|(x,y) - (x_0, y_0)\| < r_0 \leq x_0$$

and therefore $-x_0 < x - x_0 < x_0$, whence in particular $x > 0$. Similarly we check that $y > 0$, so that $(x, y) \in A$.

Continuous functions from $\mathbb{R}^n$ to $\mathbb{R}^m$

Definition 0.0.1. Let $A \subset \mathbb{R}^n$, let $f : A \to \mathbb{R}^m$, and let $\mathbf{x}_0 \in A$. We say that f is **continuous at the point** $\mathbf{x}_0$ if given any $\epsilon > 0$, there exists a $\delta > 0$ such that for any $\mathbf{x} \in A \cap B(\mathbf{x}_0, \delta)$ we have $f(\mathbf{x}) \in B(f(\mathbf{x}_0), \epsilon)$.

Definition 0.0.1 immediately implies a basic property of continuous functions. Recall that given any three sets A, B, and C and given any two functions $f : A \to B$ and $g : B \to C$, the *composition* of g with f, denoted $g \circ f$, is the map defined on A putting

$$(g \circ f)(x) = g(f(x)), \quad x \in A.$$

In order for the above definition to make sense, it is clearly enough that g be defined on the subset $f(A)$ of B.

Theorem 0.0.1. *Let $A \subset \mathbb{R}^n$, let $f : A \to \mathbb{R}^m$, and let $g : B \to \mathbb{R}^p$ with $f(A) \subset B \subset \mathbb{R}^m$. If f is continuous at $\mathbf{x}_0 \in A$ and g is continuous at $f(\mathbf{x}_0)$, then $g \circ f$ is continuous at $\mathbf{x}_0$.*

Proof. Let $\epsilon > 0$ be given. As g is continuous at $\mathbf{y}_0 = f(\mathbf{x}_0)$, there is an $r > 0$ so that for any $\mathbf{y} \in B \cap B(\mathbf{y}_0, r)$ we have $g(\mathbf{y}) \in B(g(\mathbf{y}_0), \epsilon)$. In turn, the continuity of f at $\mathbf{x}_0$ guarantees that, for some $\delta > 0$, $f(\mathbf{x}) \in B(\mathbf{y}_0, r)$ for any $\mathbf{x} \in A \cap B(\mathbf{x}_0, \delta)$. It follows that for any such $\mathbf{x}$ we have $g(f(\mathbf{x})) \in B(g(\mathbf{y}_0), \epsilon)$, whence the result. □

As a consequence of Theorem 0.0.1 and of the continuity in $\mathbb{R}^2 = \mathbb{R} \times \mathbb{R}$ of the algebraic operations $\mathbf{x} = (x, y) \to x + y$ and $\mathbf{x} = (x, y) \to xy$ (or by direct application of the definition), we obtain in particular the following theorem.

Theorem 0.0.2. *Let $A \subset \mathbb{R}^n$ and let $f, g : A \to \mathbb{R}$. If f and g are continuous at $\mathbf{x}_0 \in A$, then also $f + g$ and fg are continuous at $\mathbf{x}_0$.*

We recall the following fundamental result, which will be proved in more generality in Section 2.5 of Chapter 2. A subset $A \subset \mathbb{R}^n$ is *closed* if its complement $A^c \equiv \mathbb{R}^n \setminus A$ is open. A is said to be *bounded* if it is contained in some ball.

Theorem 0.0.3 (Weierstrass theorem). *Let $K \subset \mathbb{R}^n$ be closed and bounded and let $f : K \to \mathbb{R}$ be continuous. Then f attains its maximum and minimum values in K; that is, there exist points $\mathbf{x}_0, \mathbf{x}_1 \in K$ such that*

$$f(\mathbf{x}_0) \leq f(\mathbf{x}) \leq f(\mathbf{x}_1) \quad \forall \mathbf{x} \in K.$$

Differential calculus for functions of several variables: partial and directional derivatives

Definition 0.0.2. Let $A \subset \mathbb{R}^n$, let $f : A \to \mathbb{R}$, and let $\boldsymbol{x}_0$ be an interior point of A. Given a vector $\boldsymbol{v} \in \mathbb{R}^n$, $v \neq 0$, if the limit

$$\frac{\partial f}{\partial \boldsymbol{v}}(\boldsymbol{x}_0) \equiv \lim_{t\to 0} \frac{f(\boldsymbol{x}_0 + t\boldsymbol{v}) - f(\boldsymbol{x}_0)}{t} \tag{0.0.4}$$

exists and is finite, we call it the **directional derivative of f at x_0 along the direction** v.

Remark 0.0.1. As $\boldsymbol{x}_0$ is assumed to be interior to A, there is an $r > 0$ such that $B(\boldsymbol{x}_0, r) \subset A$. Thus, for $t \in \mathbb{R}$ such that $|t| < r/\|\boldsymbol{v}\|$, we have $\boldsymbol{x}_0 + t\boldsymbol{v} \in A$, for

$$\|\boldsymbol{x}_0 + t\boldsymbol{v} - \boldsymbol{x}_0\| = \|t\boldsymbol{v}\| = |t|\|\boldsymbol{v}\| < r.$$

This shows that the map

$$\phi_{\boldsymbol{v}} : t \to f(\boldsymbol{x}_0 + t\boldsymbol{v})$$

is well defined in the neighborhood $J_{\boldsymbol{v}} \equiv] - r/\|\boldsymbol{v}\|, r/\|\boldsymbol{v}\|[$ of $t = 0$, and (0.0.4) then shows that

$$\frac{\partial f}{\partial \boldsymbol{v}}(\boldsymbol{x}_0) = \lim_{t\to 0} \frac{\phi_{\boldsymbol{v}}(t) - \phi_{\boldsymbol{v}}(0)}{t} = \frac{d\phi_{\boldsymbol{v}}}{dt}(0) = \phi'_{\boldsymbol{v}}(0). \tag{0.0.5}$$

In other words, the directional derivative of f at $\boldsymbol{x}_0$ along the direction $\boldsymbol{v}$ is nothing but the derivative at $t = 0$ of the auxiliary function $\phi_{\boldsymbol{v}}$, which in turn represents the restriction of f to the straight line $\boldsymbol{x}_0 + t\boldsymbol{v}$ passing through the point $\boldsymbol{x}_0$ with direction $\boldsymbol{v}$.

When $\boldsymbol{v} = \boldsymbol{e}_i$, the ith unit vector of the canonical basis of $\mathbb{R}^n$, one writes $\frac{\partial f}{\partial x_i}$ rather than $\frac{\partial f}{\partial \boldsymbol{e}_i}$ and calls $\frac{\partial f}{\partial x_i}$ the ith **partial derivative of** f. Assuming that all partial derivatives of f exist at the point x_0, one puts

$$\nabla f(\boldsymbol{x}_0) = \left(\frac{\partial f}{\partial x_1}(\boldsymbol{x}_0), \ldots, \frac{\partial f}{\partial x_n}(\boldsymbol{x}_0)\right)$$

and calls $\nabla f(\boldsymbol{x}_0)$ the **gradient** of f at $\boldsymbol{x}_0$.

Differentials

Definition 0.0.3. Let $A \subset \mathbb{R}^n$, let $f : A \to \mathbb{R}$, and let $\boldsymbol{x}_0$ be an interior point of A. We say that f is **differentiable at $\boldsymbol{x}_0$** if there exists a linear map $L : \mathbb{R}^n \to \mathbb{R}$ such that $f(\boldsymbol{x}_0 + \boldsymbol{h}) = f(\boldsymbol{x}_0) + L(\boldsymbol{h}) + o(\|\boldsymbol{h}\|)$ as $\boldsymbol{h} \to \boldsymbol{0}$, that is to say, such that

$$\lim_{\boldsymbol{h}\to\boldsymbol{0}} \frac{f(\boldsymbol{x}_0 + \boldsymbol{h}) - f(\boldsymbol{x}_0) - L(\boldsymbol{h})}{\|\boldsymbol{h}\|} = \boldsymbol{0}.$$

In this case the map L (which is unique) is called the **differential of f at the point** x_0.

It is easy to check that if $n = 1$, i. e., if f is a function of one variable, then differentiability at a point x_0 is equivalent to the existence of the derivative $f'(x_0)$ at that point (and in this case, the linear map $L : \mathbb{R} \to \mathbb{R}$ indicated in Definition 0.0.3 is simply the map $h \to f'(x_0)h$). In the general case, differentiability is a stronger property than the mere existence of the partial derivatives, as shown by the next theorem (which we report without proof) and by the existence of counterexamples to the reverse implication.

Theorem 0.0.4. *Let $A \subset \mathbb{R}^n$, let $f : A \to \mathbb{R}$, and let $\boldsymbol{x}_0$ be an interior point of A. If f is differentiable at $\boldsymbol{x}_0$, then f has directional derivative at $\boldsymbol{x}_0$ along any direction $\boldsymbol{v}$, and*

$$\frac{\partial f}{\partial \boldsymbol{v}}(\boldsymbol{x}_0) = L(\boldsymbol{v}),$$

where L is the differential of f at $\boldsymbol{x}_0$. In particular, $\frac{\partial f}{\partial x_i}(\boldsymbol{x}_0) = L(\boldsymbol{e}_i)$ for $i = 1, \dots, n$, and therefore

$$L(\boldsymbol{v}) = \nabla f(\boldsymbol{x}_0) \cdot \boldsymbol{v} \quad (\boldsymbol{v} \in \mathbb{R}^n).$$

Theorem 0.0.4 tells us that the differential of f at x_0 is **represented** by the gradient vector $\nabla f(\boldsymbol{x}_0)$.

The next result we recall says that if the existence of the partial derivatives is accompanied by their **continuity**, then we regain differentiability.

Theorem 0.0.5. *Let $A \subset \mathbb{R}^n$, let $f : A \to \mathbb{R}$, and let $\boldsymbol{x}_0$ be an interior point of A. Suppose that f possesses all partial derivatives in a neighborhood of $\boldsymbol{x}_0$ and that they are all continuous at $\boldsymbol{x}_0$. Then f is differentiable at $\boldsymbol{x}_0$.*

A function f defined in an open subset A of $\mathbb{R}^n$ possessing continuous partial derivatives in the whole of A is said to be **of class C^1 in** A.

By virtue of the definitions and theorems recalled above, many important results from the differential calculus for functions of one variable can be extended to functions of several variables. We illustrate this with two important examples.

Theorem 0.0.6 (Fermat's theorem). *Let $A \subset \mathbb{R}^n$ and let $f : A \to \mathbb{R}$. Suppose that f attains a local maximum or minimum at the point $\boldsymbol{x}_0 \in A$. Then, if $\boldsymbol{x}_0$ is interior to A and f has partial derivatives at $\boldsymbol{x}_0$, we have $\nabla f(\boldsymbol{x}_0) = 0$.*

Proof. Suppose for instance that $\boldsymbol{x}_0$ is a (local) minimum point for f; this means that there is an $r > 0$ such that $B(\boldsymbol{x}_0, r) \subset A$ and

$$f(\boldsymbol{x}) \geq f(\boldsymbol{x}_0) \quad \forall \boldsymbol{x} \in B(\boldsymbol{x}_0, r).$$

Fix an index $i \in \{1, \dots, n\}$ and observe that $\boldsymbol{x}_0 + t\boldsymbol{e}_i \in B(\boldsymbol{x}_0, r)$ for any $t \in \,]-r, r[$, whence

$$h(t) \equiv f(\boldsymbol{x}_0 + t\boldsymbol{e}_i) \geq f(\boldsymbol{x}_0) = h(0)$$

for all such t, showing that the function h has a local minimum at $t = 0$.

Therefore, $h'(0) = \frac{\partial f}{\partial x_i}(\mathbf{x}_0) = 0$, and as this holds for all i the result follows. □

Theorem 0.0.7 (Mean value theorem). *Let A be an open subset of $\mathbb{R}^n$ and let $f : A \to \mathbb{R}$ be of class C^1 in A. Let $\mathbf{x}, \mathbf{y}$ be two points in A and suppose that the segment joining $\mathbf{x}$ and $\mathbf{y}$ lies in A. Then there is a point $\mathbf{z}$ on the segment such that*

$$f(\mathbf{y}) - f(\mathbf{x}) = \nabla f(\mathbf{z}) \cdot (\mathbf{y} - \mathbf{x}).$$

Proof. By the assumption on the pair $\mathbf{x}, \mathbf{y}$ we know that

$$\mathbf{x} + t(\mathbf{y} - \mathbf{x}) \in A \quad \text{for every } t \in [0, 1].$$

Now reason as before: put $\mathbf{v} = \mathbf{y} - \mathbf{x}$ and consider the function g defined in $[0, 1]$ by the equality

$$g(t) = f(\mathbf{x} + t\mathbf{v}). \tag{0.0.6}$$

We claim that g is differentiable in $[0, 1]$ and that

$$g'(t_0) = \frac{\partial f}{\partial \mathbf{v}}(\mathbf{x} + t_0\mathbf{v}) \quad \forall t_0 \in [0, 1]. \tag{0.0.7}$$

Indeed,

$$g'(t_0) = \lim_{h \to 0} \frac{g(t_0 + h) - g(t_0)}{h} = \lim_{h \to 0} \frac{f(\mathbf{x} + (t_0 + h)\mathbf{v}) - f(\mathbf{x} + t_0\mathbf{v})}{h},$$

so we obtain (0.0.7) by Definition 0.0.2 of the directional derivative and the assumption that f is of class C^1 in the whole open set A, which allows to use Theorems 0.0.4 and 0.0.5; by the former of these, we also know that

$$\frac{\partial f}{\partial \mathbf{v}}(\mathbf{x} + t_0\mathbf{v}) = \nabla f(\mathbf{x} + t_0\mathbf{v}) \cdot \mathbf{v}. \tag{0.0.8}$$

Now return to our auxiliary function g. By Lagrange's mean value theorem, we know that there is a $\hat{t} \in]0, 1[$ such that

$$g(1) - g(0) = g'(\hat{t}),$$

and if we rewrite this equality using (0.0.6), (0.0.7), and (0.0.8), we obtain

$$f(\mathbf{y}) - f(\mathbf{x}) = \nabla f(\mathbf{x} + \hat{t}(\mathbf{y} - \mathbf{x})) \cdot (\mathbf{y} - \mathbf{x}).$$

Putting $\mathbf{z} = \mathbf{x} + \hat{t}(\mathbf{y} - \mathbf{x})$, this ends the proof of Theorem 0.0.7. □

Notations

List of most frequently used notations

- The symbols **x**, **y** (or $\boldsymbol{x}, \boldsymbol{y}$) are reserved to vectors of $\mathbb{R}^n (n > 1)$.
- I, J will always denote intervals (bounded or not, closed or not) in $\mathbb{R}$.
- The symbols s, t are reserved to real numbers, and in particular they are used in Chapter 1 to denote the independent variable in ODEs.
- $x'(t)$, $\frac{dx}{dt}(t)$, and $(Dx)(t)$ denote the derivative of the function $x = x(t)$ of the real variable t.
- $\int x(t)dt$ (indefinite integral of the function $x = x(t)$) denotes any primitive of the function $x = x(t)$ (that is, any function F such that $F' = x$).
- $A \times B \equiv \{(a, b) : a \in A, b \in B\}$ is the Cartesian product of the sets A, B.
- $B(x_0, r) = \{x \in X : d(x, x_0) < r\}$ is the (open) ball of center $x_0 \in X$ in a metric space (X, d).
- Ω will usually denote a bounded open set in $\mathbb{R}^n$.
- $C(A, \mathbb{R}^m)$, with $A \subset \mathbb{R}^n$, denotes the set of all continuous functions from A to $\mathbb{R}^m$. In case $m = 1$, we write $C(A)$ rather than $C(A, \mathbb{R})$. A similar proviso holds for other function spaces too.
- $\frac{\partial f}{\partial x_i}(\boldsymbol{x}_0), f_{x_i}(\boldsymbol{x}_0)$ denote the partial derivative of $f = f(\boldsymbol{x}) = f(x_1, \ldots, x_n)$ with respect to x_i at the point $\boldsymbol{x}_0$.
- $\nabla f(\boldsymbol{x}_0) = (\frac{\partial f}{\partial x_1}(\boldsymbol{x}_0), \ldots, \frac{\partial f}{\partial x_n}(\boldsymbol{x}_0))$ is the gradient of $f = f(\boldsymbol{x}) = f(x_1, \ldots, x_n)$ at $\boldsymbol{x}_0$.
- $\Delta f = \frac{\partial^2 f}{\partial x_1^2} + \cdots + \frac{\partial^2 f}{\partial x_n^2}$ is the Laplacian of f.

https://doi.org/10.1515/9783111302522-203

1 Ordinary differential equations

Introduction

This chapter begins with the well-known formula (see, e. g., Apostol's book *Calculus* [4])

$$x(t) = c\, e^{\int a(t)\,dt} + e^{\int a(t)\,dt} \int e^{-\int a(t)\,dt} b(t)dt, \tag{1.0.1}$$

giving the explicit solutions of the linear equation $x' = a(t)x + b(t)$, and one of its targets is to lead the reader to the extension of (1.0.1) to first-order linear systems such as

$$X' = A(t)X + B(t), \tag{1.0.2}$$

where A, B are, respectively, an $n \times n$ matrix and an n-column vector with real continuous elements. As will be clear from the discussion in Section 1.3, this generalization is based on the fusion (if we can say so) between some well-known and basic facts from linear algebra on the one hand and the (global) existence theorem for the Cauchy problem attached to (1.0.2) (see Lemma 1.3.1) on the other hand.

In this presentation, two intermediate steps need to be made for the described extension of (1.0.1) to (1.0.2). The first consists in passing from the linear equation $x' = a(t)x + b(t)$ to the general first-order equation (in normal form)

$$x' = f(t, x). \tag{1.0.3}$$

We do this in Section 1.1 with a quick look at the main existence and uniqueness theorems concerning the initial value problem (IVP) for (1.0.3); practically no proofs will be given for the statements involved, and we refer since this very beginning to for instance the books of Hale [5] or Walter [1] for an adequate treatment of this topic. The second intermediate step consists of course in passing from the first-order equation (1.0.3) to first-order general systems $X' = F(t, X)$, and this is sketched in a straightforward way in Section 1.2.

Section 1.4 deals with the important case in which the matrix A in (1.0.2) is independent from t (linear systems with constant coefficients) and gives the opportunity to discuss a nice and conceptually important extension of the familiar Taylor series expansion

$$e^x = 1 + x + \frac{x^2}{2!} + \cdots =$$

of the exponential function of the real variable x to the case where x is replaced by an n-by-n matrix A.

The final Sections 1.5 and 1.6 of this chapter deal with higher-order differential equations, with the declared target of gaining some familiarity with second-order linear equations. Giving special importance to the second-order equation

https://doi.org/10.1515/9783111302522-001

$$x'' = f(t, x, x') \tag{1.0.4}$$

is justified not only by its importance in physics coming from the fundamental principle of dynamics

$$\boldsymbol{F} = m\boldsymbol{a}$$

and thus from the classical paradigm stating the uniqueness of the motion of a particle with given initial position and velocity, but also by the fact that (1.0.4) are the simplest to examine from the point of view of *boundary value problems* (BVPs), in which the initial conditions on x and x' at a given point t_0 belonging to the interval $I = [a, b]$ (assuming f is defined on $I \times \mathbb{R}^2$) are replaced by conditions involving x and/or x' at the endpoints a and b of the interval. Such kind of problems (i) yield a good starting point for the study of nonlinear operator equations and (ii) give rise to the classical theory of Sturm and Liouville for linear equations, which will be cited with some details in the Additions to Chapter 3. Finally, they serve as an introduction to the much more difficult BVPs for second-order partial differential equations (PDEs), in particular those of mathematical physics, which we shall synthetically discuss in Chapter 4.

1.1 Ordinary differential equations (ODEs) of the first order

Example 1.1.1. The first-order linear equation

$$x' = a(t)x + b(t), \tag{1.1.1}$$

with coefficients $a, b \in C(I)$, can be solved by means of the explicit formula

$$x(t) = c\, e^{\int a(t)\,dt} + e^{\int a(t)\,dt} \int e^{-\int a(t)\,dt} b(t)dt, \tag{1.1.2}$$

where for $f \in C(I)$, the symbol $\int f(t)\,dt$ denotes an arbitrarily chosen primitive of f in I. For instance, the equation

$$x' = -\frac{x}{t} + 1, \quad t \in]0, +\infty[\equiv I,$$

has the solutions $x(t) = c/t + t/2, c \in \mathbb{R}$.

Example 1.1.2. Some simple nonlinear equations can also be solved explicitly. For instance, consider

$$x' = a(t)h(x),$$

where $h \in C(B)$, $B \subset \mathbb{R}$. If along a solution $x(t)$ we have $h(x(t)) \neq 0$, then $x'(t)/h(x(t)) = a(t)$, and integrating both members of this equality gives $x(t)$ itself if one is able to find a primitive of $1/h$. For instance, the equations

$$\text{(i) } x' = 3x^{\frac{2}{3}}, \quad \text{(ii) } x' = -2tx^2, \quad \text{(iii) } x' = x(1-x)$$

can be solved to yield respectively

$$\text{(i) } x(t) = (t+k)^3, \quad \text{(ii) } x(t) = \frac{1}{t^2+k}, \quad \text{(iii) } x(t) = \frac{e^t}{e^t+k},$$

with $k \in \mathbb{R}$. The three equations all have in addition the solution $x \equiv 0$; moreover, (iii) has a second "trivial" solution, namely, $x \equiv 1$.

Consider now a first-order differential equation in general (normal) form:

$$x' = f(t,x), \tag{1.1.3}$$

where $f = f(t,x)$ is a real-valued function defined in a subset A of $\mathbb{R}^2$. To study (1.1.3) we first need to define precisely what is a *solution* of it.

Definition 1.1.1. A *solution* of the differential equation (1.1.3) is a function $u = u(t)$, defined in some interval $J = J_u \subset \mathbb{R}$, such that:
(a) u is differentiable in J,
(b) $(t, u(t)) \in A$ for all $t \in J$, and
(c) $u'(t) = f(t, u(t))$ for all $t \in J$.

Exercise 1.1.1. Prove that if f is continuous on its domain A, then any solution of (1.1.3) is of class C^1 on its interval of definition.

Remark on the notations. In order to emphasize and clarify the concept of solution, we have employed in Definition 1.1.1 a different symbol for the unknown (x) of the differential equation and for a solution (u) of it. Of course, this is simply a matter of taste in the use of notations, and throughout these Notes we will ourselves use the same symbol with both meanings (as already done in the examples displayed above).

These examples suggest that we have to expect infinitely many solutions of an equation like (1.1.3). A fundamental remark is that in Example 1.1.1, there is *exactly one* solution satisfying a given "initial" condition of the form $x(t_0) = x_0$: just impose that condition in (1.1.2) to obtain uniquely c. One can check that it is the same for equations (ii) and (iii) in Example 1.1.2, but it is *not* the same for (i) if we take the initial condition $x(t_0) = 0$, for we have in this case the two different solutions $u_1(t) = (t - t_0)^3$ and $u_2(t) = 0$.

Driven by these remarks, we shall study from now on the *initial value problem* (IVP), also called *Cauchy problem*, for the differential equation (1.1.3), which is written

$$\begin{cases} x' = f(t,x) \\ x(t_0) = x_0 \end{cases} \tag{1.1.4}$$

and consists in finding a solution u of the differential equation, defined in an interval J containing the point t_0 and such that $u(t_0) = x_0$. Of course the point $(t_0, x_0) \in A$, and we shall look for assumptions on f which guarantee for *any* such initial point the existence and uniqueness of a solution to (1.1.4) at least in a "small" interval around t_0.

The first assumption one makes about f is that it be *continuous* on its domain A. For such f, consider the *integral equation* (relative to the point $(t_0, x_0) \in A$)

$$x = x_0 + \int_{t_0}^{t} f(s,x)\, ds \tag{1.1.5}$$

and define a *solution of* (1.1.5) to be a continuous function u, defined on some interval J containing t_0 and satisfying the equation in J, meaning that $(t, u(t)) \in A$ for $t \in J$ and

$$u(t) = x_0 + \int_{t_0}^{t} f(s,u(s))\, ds \quad \forall t \in J. \tag{1.1.6}$$

Lemma 1.1.1. *Suppose that $f : A \to \mathbb{R}$ is continuous. Then the IVP* (1.1.4) *is* equivalent *to the integral equation* (1.1.5), *in the sense that any solution of* (1.1.4) *is a solution of* (1.1.5) *in the same interval, and vice versa.*

Proof. This is a mere consequence of the definitions, Exercise 1.1.1, and the fundamental theorem of calculus.

Consider now the special case in which A is a strip in the plane parallel to the x-axis, i. e., $A = I \times \mathbb{R}$ with $I \subset \mathbb{R}$ an interval. This geometrical assumption helps to make it clear that the solutions of (1.1.5) – that is, of (1.1.4) – are precisely the fixed points of a particular map F. Indeed, first note that given any map $u : J \to \mathbb{R}$, the condition $(t, u(t)) \in A = I \times \mathbb{R}$ – required to give meaning to $f(t, u(t))$ – simply means that $J \subset I$. Moreover, by (1.1.6) we have the following lemma. □

Lemma 1.1.2. *Suppose that $f : I \times \mathbb{R} \to \mathbb{R}$ is continuous. Let $(t_0, x_0) \in I \times \mathbb{R}$ and let J be an interval containing t_0. Finally, let $F : C(J) \to C(J)$ be the mapping defined as follows: for $u \in C(J)$,*

$$F(u)(t) = x_0 + \int_{t_0}^{t} f(s,u(s))\, ds, \quad t \in J. \tag{1.1.7}$$

Then u is a solution of (1.1.5) *in the interval J if and only if $F(u) = u$.*

Remark 1.1.1. That the map F defined in (1.1.7) operates *in* $C(J)$ follows from the continuity of f, which was already employed before. F is directly related to the map N_f

acting in $C(J)$ and defined by the equality $N_f(u)(t) = f(t, u(t))$; this is sometimes called the *Nemytskii operator* induced by the function f. The study of Nemytskii's operator is fundamental when dealing with nonlinear problems. To see explicitly how F works, take for instance $J = I = \mathbb{R}$, $f(t,x) = x^3$, $(t_0, x_0) = (0,0)$. Then

$$F(u)(t) = \int_0^t u^3(s)\, ds, \quad t \in \mathbb{R}.$$

If we take for instance $u(t) = \sin t$, then it is easily checked that

$$F(u)(t) = -\cos t + \frac{(\cos t)^3}{3} + \frac{2}{3}.$$

Local existence and uniqueness of solutions to the IVP

In the remaining of this section, we state without proof some basic existence and uniqueness results for solutions of the IVP (1.1.4); proofs of these statements can be found, for instance, in the book [1]. First, some more definitions are needed.

Definition 1.1.2. A function $f : A \subset \mathbb{R}^2 \to \mathbb{R}$ is said to be *Lipschitzian with respect to the second variable in A* if there exists a constant $L > 0$ such that

$$|f(t,x) - f(t,y)| \le L|x - y| \quad \forall (t,x), (t,y) \in A. \tag{1.1.8}$$

Definition 1.1.3. Let A be an open subset of $\mathbb{R}^2$. A map $f : A \subset \mathbb{R}^2 \to \mathbb{R}$ is said to be *locally Lipschitzian with respect to the second variable in A* if each point $(t_0, x_0) \in A$ has a neighborhood $U = U(t_0, x_0) \subset A$ in which f is Lipschitzian.

Here we provide the standard form of the local existence and uniqueness principle for solutions of (1.1.4).

Theorem 1.1.1. *Let $f : A \to \mathbb{R}$ with A an open subset of $\mathbb{R}^2$. Assume that:*
(a) *f is continuous in A;*
(b) *f is locally Lipschitzian with respect to the second variable in A.*

Then given any $(t_0, x_0) \in A$, there exists a neighborhood I_0 of t_0 such that (1.1.4) *has a unique solution defined in I_0.*

Example 1.1.3. The system

$$\begin{cases} x' = \sin(a(t)x) \\ x(t_0) = x_0, \end{cases} \tag{1.1.9}$$

where a is continuous on an interval I, can be solved uniquely for each point $x_0 \in \mathbb{R}$ and near each point $t_0 \in I$. Indeed, letting I_0 be a closed, bounded neighborhood of t_0 contained in I, we have

$$|\sin(a(t)x) - \sin(a(t)y)| \le |a(t)||x - y| \le L|x - y|,$$

with $L = \max_{t \in I_0} |a(t)|$.

In practice, to check the local Lipschitz condition (b) in Theorem 1.1.1 is an easy task when (as in the example above) $f = f(t, x)$ has a continuous partial derivative with respect to x in A. Indeed, recall that if g is a differentiable function defined on an interval $I \subset \mathbb{R}$, then given any two points x and y in I we have $g(x) - g(y) = g'(z)(x - y)$ for some z between x and y (mean value theorem). Then we have the following proposition.

Proposition 1.1.1. *Suppose that $f : A \to \mathbb{R}$ (with A open in $\mathbb{R}^2$) satisfies the following conditions:*
(i) *the partial derivative $\frac{\partial f}{\partial x}(t, x)$ exists at each point $(t, x) \in A$;*
(ii) *the function $\frac{\partial f}{\partial x} : A \to \mathbb{R}$ is continuous in A.*

Then f is locally Lipschitzian with respect to x in A.

Proof. Pick a point $(t_0, x_0) \in A$ and let $R = [a, b] \times [c, d]$ be a rectangular neighborhood of (t_0, x_0) contained in A. Then for any $(t, x), (t, y) \in R$ there exists a z between x and y such that

$$|f(t, x) - f(t, y)| = \left| \frac{\partial f}{\partial x}(t, z)(x - y) \right| \le K|x - y|,$$

with $K = \max_{(t,x) \in R} |\frac{\partial f}{\partial x}(t, x)|$; K is well defined because of (ii) and the Weierstrass theorem (Theorem 0.0.3) on continuous functions over compact sets. □

Example 1.1.4. The equations

$$x' = te^x - \log x, \quad x' = \frac{\arctan(tx)}{\log(1 - (t^2 + x^2))}$$

can be solved uniquely near each point (t_0, x_0) belonging respectively to the half-space $\{(t, x) : x > 0\}$ or to the punctured ball $\{(t, x) \in \mathbb{R}^2 : 0 < t^2 + x^2 < 1\}$, which both are open subsets of $\mathbb{R}^2$.

Exercise 1.1.2. With reference to (i) of Example 1.1.2, show that the function $f : \mathbb{R} \times \mathbb{R} \to \mathbb{R}$ defined putting

$$f(t, x) = x^{\frac{2}{3}}$$

is *not* locally Lipschitzian.

Continuation of solutions, maximal solutions, global solutions

Definition 1.1.4. Let u and v be two solutions of $x' = f(t, x)$ in the intervals I_u and I_v, respectively. We say that v is a *continuation of u* if $I_u \subset I_v$ and $v = u$ in I_u. The continuation is said to be *proper* if $I_u \subset I_v$ properly. A solution is said to be *maximal* if it has no proper continuation.

Lemma 1.1.3. *Suppose that the assumptions of Theorem* 1.1.1 *are satisfied and let u and v be two solutions of $x' = f(t,x)$ defined in the same interval J. If $u(\hat{t}) = v(\hat{t})$ for some $\hat{t} \in J$, then $u(t) = v(t)$ for all $t \in J$.*

Theorem 1.1.2. *Suppose that the assumptions of Theorem* 1.1.1 *are satisfied. Then any solution of $x' = f(t,x)$ has a unique maximal continuation.*

Now let $(t_0, x_0) \in A$ and let u be the "local" solution of the corresponding IVP (1.1.4), whose existence is granted by Theorem 1.1.1 and defined in a neighborhood I_0 of t_0. The theorem just stated ensures that u_0 can be uniquely continued to a maximal interval containing I_0. Therefore, we have the following theorem.

Theorem 1.1.3. *Suppose that the assumptions of Theorem* 1.1.1 *are satisfied. Then for any $(t_0, x_0) \in A$, the IVP* (1.1.4) *has a* unique *maximal solution.*

Question: where are the (maximal) solutions of an ODE defined?
On the basis of Theorem 1.1.2, when speaking of a *solution* of $x' = f(t,x)$, one always means a *maximal* solution. Now, the question written above has no definite meaning in general, but it does when $A = I \times \mathbb{R}$, in which case it is natural to ask whether the solutions are defined on the whole of I.

Definition 1.1.5. Let $f : I \times \mathbb{R} \to \mathbb{R}$. A solution of $x' = f(t,x)$ is said to be *global* if it is defined on the whole of I.

For instance, looking at the three equations appearing in Example 1.1.2 (in all of which $I = \mathbb{R}$), we see that the solutions of (i) are all global, while for (ii) and (iii) some of them are global and some others (precisely, those for $k < 0$) are not. Moreover, we can consider for instance the equation $x' = x^2$, having the global (trivial) solution $u \equiv 0$ and the solutions

$$u(t) = -\frac{1}{t+k},$$

none of which is global. It is therefore natural to ask which additional conditions one has to impose on f in order to guarantee that *all* solutions of $x' = f(t,x)$ are global.

Definition 1.1.6. A map $f : I \times \mathbb{R} \to \mathbb{R}$ is said to be *sublinear* with respect to the second variable if there exist $\alpha, \beta \in C(I)$ such that

$$|f(t,x)| \le \alpha(t)|x| + \beta(t) \tag{1.1.10}$$

for all $t \in I$ and $x \in \mathbb{R}$.

Examples.

(i) If $f(t,x) = a(t)x + b(t)$ with $a, b \in C(I)$, then f is sublinear.
(ii) If $f(t,x) = \sqrt{|x|}$, then f is sublinear since $\sqrt{|x|} \le |x| + 1$ for all $x \in \mathbb{R}$.
(iii) If $f(t,x) = x^2$, it is easy to check that f is *not* sublinear.

Theorem 1.1.4. *Let $f : I \times \mathbb{R} \to \mathbb{R}$ with I an open interval in $\mathbb{R}$. Assume that:*
(a) *f is continuous in $I \times \mathbb{R}$;*
(b) *f is locally Lipschitzian with respect to the second variable in $I \times \mathbb{R}$;*
(c) *f is sublinear with respect to the second variable in $I \times \mathbb{R}$.*

Then given any $(t_0, x_0) \in I \times \mathbb{R}$, the maximal solution of (1.1.4) *is defined on I. In other words, for any (t_0, x_0) the IVP* (1.1.4) *has a unique* global *solution.*

1.2 Systems of first-order differential equations

A system of two differential equations of the first order in the two unknowns x, y has the form

$$\begin{cases} x' = f(t,x,y) \\ y' = g(t,x,y), \end{cases} \tag{1.2.1}$$

where $f, g : A \subset \mathbb{R}^3 \to \mathbb{R}$ are given functions. For instance,

$$\begin{cases} x' = x(a - by) \\ y' = y(-c + dx), \end{cases} \tag{1.2.2}$$

where a, b, c, d are all positive constants, is the *Lotka–Volterra system* modeling prey–predator competition. In general, a system of n first-order differential equations has the form

$$\begin{cases} x_1' = f_1(t, x_1, x_2, \ldots, x_n) \\ x_2' = f_2(t, x_1, x_2, \ldots, x_n) \\ \ldots \\ x_n' = f_n(t, x_1, x_2, \ldots, x_n), \end{cases} \tag{1.2.3}$$

where $f_1, \ldots, f_n$ are given real-valued functions defined in a subset A of $\mathbb{R}^{n+1}$ and can be written in vector form as

$$X' = F(t, X), \tag{1.2.4}$$

where $X = (x_1, x_2, \ldots x_n) \in \mathbb{R}^n$ and $F = F(t,X) : A \to \mathbb{R}^n$ is the vector-valued function, defined in A, whose component functions are $(f_1, f_2, \ldots, f_n)$.

The vector form (1.2.4) of the system (1.2.3) allows for an almost immediate extension of the definitions and results already seen for the scalar equation $x' = f(t,x)$. Here, we just resume the main points of this extension and refer again to [1] for details and proofs.

A. Solution of (1.2.4)
A *solution* of (1.2.4) is a function U, defined in an interval $J_U \subset \mathbb{R}$, that is differentiable in J_U such that, for all $t \in J_U$,

$$(t, U(t)) \in A \quad \text{and} \quad U'(t) = F(t, U(t)).$$

B. The IVP for (1.2.4)
The IVP for (1.2.4) is written

$$\begin{cases} X' = F(t,X) \\ X(t_0) = X_0, \end{cases} \tag{1.2.5}$$

where $(t_0, X_0) \in A$, and consists in finding a solution U of (1.2.4) defined in a neighborhood of t_0 such that $U(t_0) = X_0$.

C. Lipschitz and locally Lipschitz functions $F : A \subset \mathbb{R}^{n+1} \to \mathbb{R}^m$
Definition 1.2.1. Let A be an open subset of $\mathbb{R}^{n+1}$. A map $F = F(t,X) : A \to \mathbb{R}^m$ is said to be *locally Lipschitzian with respect to X in A* if each point $(t_0, X_0) \in A$ has a neighborhood $U = U(t_0, X_0) \subset A$ where F is Lipschitzian with respect to X, that is, it satisfies the inequality

$$\|F(t,X) - F(t,Y)\| \le L\|X - Y\| \tag{1.2.6}$$

for some $L > 0$ and for all $(t,X), (t,Y) \in U$.

In (1.2.6), the symbol $\|.\|$ stands for the Euclidean norm in $\mathbb{R}^n$:

$$\|X\| = \sqrt{\sum_{i=1}^{n} x_i^2} \quad \text{if } X = (x_1, \dots, x_n).$$

D. Local existence and uniqueness of solutions of the IVP (1.2.5)
Theorem 1.2.1. *Let $F : A \to \mathbb{R}^n$ with A an open subset of $\mathbb{R}^{n+1}$. Assume that:*
(a) *F is continuous in A;*
(b) *F is locally Lipschitzian with respect to the second variable X in A.*

Then given any $(t_0, X_0) \in A$, there exists a neighborhood I_0 of t_0 such that (1.2.5) *has a unique solution defined in I_0.*

The following propositions are helpful to check that condition (b) in Theorem 1.2.1 is satisfied.

Proposition 1.2.1. *$F = (f_1, \dots, f_m) : A \to \mathbb{R}^m$ is Lipschitz (locally Lipschitz) in $A \subset \mathbb{R}^{n+1}$ if and only if for each $i = 1, \dots, m$, $f_i : A \to \mathbb{R}$ is Lipschitz (locally Lipschitz) in A.*

Proposition 1.2.2. *$f : A \to \mathbb{R}$ is locally Lipschitz in $A \subset \mathbb{R}^{n+1}$ provided that for each $i = 1, \dots, n$, the partial derivative $\frac{\partial f}{\partial x_i}$ exists and is continuous in A.*

Example 1.2.1. Given any $t_0 \neq 0$ and any $x_0, y_0 \in \mathbb{R}$, the system

$$\begin{cases} x' = x \log y \equiv f(t, x, y) \\ y' = (x^2 + y^2)/t \equiv g(t, x, y) \\ x(t_0) = x_0, \quad y(t_0) = y_0 \end{cases} \tag{1.2.7}$$

has a unique solution defined in an appropriately small neighborhood of t_0, for the functions f and g are plainly continuous in their (open) domain of definition $A \equiv (\mathbb{R} \setminus \{0\}) \times \mathbb{R} \times]0, +\infty[$, and so are the partial derivatives

$$\frac{\partial f}{\partial x}(x, y) = \log y, \quad \frac{\partial f}{\partial y}(x, y) = \frac{x}{y}, \quad \frac{\partial g}{\partial x}(x, y) = \frac{2x}{t}, \quad \frac{\partial g}{\partial y}(x, y) = \frac{2y}{t}.$$

In fact, in this simple example f, g are of class C^1 in A. The same can be said for the system

$$\begin{cases} x' = tx \sin y \\ y' = \frac{1}{\sqrt{1-t^2}} \frac{y}{1+x^2} \\ x(t_0) = x_0, \quad y(t_0) = y_0. \end{cases} \tag{1.2.8}$$

E. Global solutions of (1.2.4) **(when $A = I \times \mathbb{R}^n$)**

Definition 1.2.2. A map $F : I \times \mathbb{R}^n \to \mathbb{R}^m$ is said to be *sublinear* with respect to the second variable if there exist $\alpha, \beta \in C(I)$ such that

$$\|F(t, X)\| \leq \alpha(t)\|X\| + \beta(t) \tag{1.2.9}$$

for all $t \in I$ and $X \in \mathbb{R}^n$.

As in Proposition 1.2.1, $F = (f_1, \dots, f_m) : A \to \mathbb{R}^m$ is sublinear if and only if each f_i is sublinear. With these definitions for the vector case at hand, we can state the general form of the global existence and uniqueness theorem.

Theorem 1.2.2. *Let $F : I \times \mathbb{R}^n \to \mathbb{R}^n$ with I an open interval in $\mathbb{R}$. If F is continuous, locally Lipschitzian with respect to X and sublinear with respect to X in $I \times \mathbb{R}^n$, then given any $(t_0, X_0) \in I \times \mathbb{R}^n$, the IVP* (1.2.5) *has a unique solution defined on all of I.*

For instance, the system (1.2.8) has a unique solution defined on $]-1,1[$; likewise, the problem

$$\begin{cases} x' = tx \arctan xy + 1 \\ y' = \frac{1}{t}\frac{xy}{1+x^2} + e^t x \\ x(1) = 0, \quad y(1) = 1 \end{cases} \tag{1.2.10}$$

has a unique solution defined on $]0,+\infty[$, for the inequality $|x|/1+x^2 \le 1$ holding for all $x \in \mathbb{R}$ shows that the right-hand member of the second equation in (1.2.10) is sublinear in each of the strips $]0,+\infty[\times \mathbb{R}^2$, $]-\infty,0[\times \mathbb{R}^2$.

Exercise 1.2.1. Show that the system

$$\begin{cases} x' = tx \sin y + 1 \equiv f(t,x,y) \\ y' = \frac{\sqrt{x^2+y^2}}{t} \equiv g(t,x,y) \\ x(1) = 0, \quad y(1) = 1 \end{cases} \tag{1.2.11}$$

has a unique solution defined on $]0,+\infty[$. In this example, the function g does not have partial derivatives in its domain of definition $A = (\mathbb{R}\setminus\{0\})\times\mathbb{R}^2$, but is nonetheless locally Lipschitzian with respect to $X = (x,y)$ in A because of the inequality

$$\big|\|X\| - \|Y\|\big| \le \|X - Y\|, \tag{1.2.12}$$

which holds for any norm by virtue of the triangle inequality.

Remark 1.2.1. The definitions and results given in Section 1.1 about the continuation of solutions and related questions remain valid also for systems of ODEs. In particular, it follows that when $F : I \times \mathbb{R}^n \to \mathbb{R}^n$ satisfies the assumptions of Theorem 1.2.2, then *every* (maximal) solution of the system $X' = F(t,X)$ is global. Indeed, any solution $U : J_U \to \mathbb{R}$ of this system is in particular a solution of the IVP (1.2.5), with t_0 any point of J_U and $X_0 = U(t_0)$. Now if we denote with $Z : I \to \mathbb{R}^n$ the global solution of the IVP (1.2.5) given by Theorem 1.2.2, it follows first (from Lemma 1.1.3) that Z is a continuation of U and then that $J_U = I$, for otherwise U would not be maximal.

1.3 Linear systems of first-order ODEs

Consider now the following special form of (1.2.1):

$$\begin{cases} x' = a(t)x + b(t)y + c(t) \\ y' = d(t)x + e(t)y + f(t), \end{cases} \tag{1.3.1}$$

where a, b, c, d, e, f are real-vaued functions defined in an interval I, called the **coefficients** of the system (1.3.1). In general, a **linear** system of n first-order differential equations has the form

$$\begin{cases} x_1' = a_{11}(t)x_1 + a_{12}(t)x_2 + \cdots + a_{1n}(t)x_n + b_1(t) \\ x_2' = a_{21}(t)x_1 + a_{22}(t)x_2 + \cdots + a_{2n}(t)x_n + b_2(t) \\ \dots \\ x_n' = a_{n1}(t)x_1 + a_{n2}(t)x_2 + \cdots + a_{nn}(t)x_n + b_n(t). \end{cases} \tag{1.3.2}$$

Putting

$$X = \begin{pmatrix} x_1 \\ x_2 \\ \vdots \\ x_n \end{pmatrix}, \quad A = \begin{pmatrix} a_{11} & \dots & a_{1n} \\ a_{21} & \dots & a_{2n} \\ \vdots & \dots & \vdots \\ a_{n1} & \dots & a_{nn} \end{pmatrix}, \quad B = \begin{pmatrix} x_1 \\ x_2 \\ \dots \\ x_n \end{pmatrix}, \tag{1.3.3}$$

(1.3.2) can be written

$$X' = A(t)X + B(t). \tag{1.3.4}$$

Consider the **homogeneous system** associated with (1.3.4):

$$X' = A(t)X. \tag{1.3.5}$$

Also, consider the **initial value problem (IVP)** associated with (1.3.4):

$$\begin{cases} X' = A(t)X + B(t) \\ X(t_0) = X_0. \end{cases} \tag{1.3.6}$$

Lemma 1.3.1. *If the coefficient functions a_{ij} and b_i are* ***continuous*** *on I, then for any $(t_0, X_0) \in I \times \mathbb{R}^n$ the maximal solution of* (1.3.6) *is defined on all of I. In other words, for any $(t_0, X_0) \in I \times \mathbb{R}^n$, the IVP* (1.3.6) *has a unique* ***global*** *solution.*

Proof. We can apply Theorem 1.2.2 because with $F(t,X) = A(t)X+B(t)$, it is easily checked that F is:
- continuous,
- locally Lipschitzian,
- sublinear.

Indeed, the ith ($i = 1, \dots, n$) component f_i of F has the explicit form

$$fi(t,X) = f_i(t, x_1, \dots, x_n) = a_{i1}(t)x_1 + a_{i2}(t)x_2 + \cdots + a_{in}(t)x_n + b_i(t),$$

so that the continuity of f_i is evident from its special dependence on X and the continuity of the coefficients $a_{i,j}$ and b_i; as to local Lipschitzianity, just observe that

$$\frac{\partial f_i}{\partial x_j}(t,X) = a_{ij}(t) \quad (t,X) \in I \times \mathbb{R}^n, \quad i,j = 1, \dots, n.$$

Finally, the sublinearity of f_i follows from the inequality

$$|f_i(t,X)| \le \sum_{j=1}^{n} |a_{ij}(t)x_j + b_i(t)| \le \alpha(t)\|X\| + \beta(t),$$

where $\alpha(t) = \sum_{j=1}^{n} |a_{ij}(t)|$ and $\beta(t) = |b_i(t)|$. □

Remark 1.3.1. It follows from Lemma 1.3.1 (and Exercise 1.1.1!) that every solution of (1.3.4) belongs to the vector space $C^1(I,\mathbb{R}^n)$ of all $\mathbb{R}^n$-valued functions defined on I and continuous on I together with their first derivative.

We will now see what special properties the solution sets of (1.3.4) and of (1.3.5) have.

Lemma 1.3.2. *Let S and S_0 be respectively the set of all solutions of* (1.3.4) *and the set of all solutions of the homogeneous system* (1.3.5), *or in symbols,*

$$S = \{U \in C^1(I,\mathbb{R}^n) : U'(t) = A(t)U(t) + B(t), t \in I\}, \tag{1.3.7}$$

$$S_0 = \{U \in C^1(I,\mathbb{R}^n) : U'(t) = A(t)U(t), t \in I\}. \tag{1.3.8}$$

Then:
(i) *S_0 is a vector subspace of $C^1(I,\mathbb{R}^n)$;*
(ii) *$S = S_0 + X_0 \equiv \{U + X_0 : U \in S_0\}$, where $X_0 \in S$ is arbitrarily fixed.*

Proof. Let T be the map defined putting, for $U \in C^1(I,\mathbb{R}^n)$,

$$T(U)(t) = U'(t) - A(t)U(t), \quad t \in I.$$

It is easy to check that T is a **linear** map of $E \equiv C^1(I,\mathbb{R}^n)$ into $F \equiv C(I,\mathbb{R}^n)$. Now since

$$S_0 = \{U \in E : T(U) = 0\},$$

(i) follows immediately (S_0 is just the kernel Ker T of T). On the other hand,

$$S = \{U \in E : T(U) = B\},$$

so that statement (ii) is another elementary fact from linear algebra. □

Proposition 1.3.1. *Let E, F be vector spaces and let T be a linear map of E into F. If the equation*

$$Tx = y \quad (y \in F) \tag{1.3.9}$$

has a solution x_0, then the set $S_y \equiv \{x \in E : Tx = y\}$ of all solutions of (1.3.9) *is given by*

$$S_y = \operatorname{Ker} T + x_0.$$

Proof. This is left as an exercise.

In brief, the statements of Lemma 1.3.2 are straightforward consequences of the linearity of the system (1.3.4). This same fact, used in conjunction with the existence and uniqueness of solutions of the IVP for first-order linear systems, stated before for convenience as Lemma 1.3.1, implies other and fundamental properties of the solution sets for equations (1.3.4) and (1.3.5). To fully appreciate this statement, let us recall formally a second fact from linear algebra. □

Proposition 1.3.2. *Let E, F be vector spaces and let T be a linear map of E into F. If T is* bijective *(i. e., it is an isomorphism) and E has dimension n, then also F has dimension n.*

Proof. This is left as an exercise. □

Theorem 1.3.1. *The set of all solutions of the homogeneous system* (1.3.5) *is a vector subspace **of dimension** n of $C^1(I, \mathbb{R}^n)$.*

Proof. Fix a point $t_0 \in I$ and consider the map H_{t_0} of $C^1(I, \mathbb{R}^n)$ into $\mathbb{R}^n$ defined as follows for every $U \in C^1(I, \mathbb{R}^n)$:

$$H_{t_0}(U) = U(t_0).$$

H_{t_0} is evidently linear; we could call H_{t_0} the *evaluation map at* t_0. Consider now the restriction K_{t_0} of H_{t_0} to S_0: we claim that K_{t_0} is a *bijective* map of S_0 onto $\mathbb{R}^n$. Indeed, let $X_0 \in \mathbb{R}^n$; by the existence and uniqueness lemma, Lemma 1.3.1 (used with $B = 0$), there is a unique $U_0 \in C^1(I, \mathbb{R}^n)$ that is a solution of the IVP

$$\begin{cases} X' = A(t)X \\ X(t_0) = X_0. \end{cases} \tag{1.3.10}$$

In other words, there is a unique $U_0 \in S_0$ such that $K_{t_0}(U) = U(t_0) = X_0$, so that K_{t_0} is in fact an isomorphism of S_0 onto $\mathbb{R}^n$. The statement in Theorem 1.3.1 thus follows from Proposition 1.3.2 – or, more literally, from the equivalent statement in which it is assumed that the target space F has dimension n. □

Fundamental system of solutions of (1.3.5). Fundamental matrix

Definition 1.3.1. A *fundamental system of solutions* of (1.3.5) is any set of n linearly independent solutions of (1.3.5), or equivalently (in view of Theorem 1.3.1) any basis of the vector space S_0 defined in (1.3.8).

It follows that if $V_1, V_2, \ldots, V_n$ are n linearly independent solutions of (1.3.5), then any solution U of (1.3.5) can be written in a unique way as a linear combination

$$U = c_1 V_1 + \cdots c_n V_n \quad (c_i \in \mathbb{R}). \tag{1.3.11}$$

The following proposition is useful when checking the linear independence (l. i.) of n solutions of (1.3.5).

Proposition 1.3.3. *Let $V_1, V_2, \ldots, V_n$ be n solutions of* (1.3.5). *The following statements are equivalent:*

(i) *there exists a* $\hat{t} \in I$ *such that* $V_1(\hat{t}), V_2(\hat{t}), \dots, V_n(\hat{t})$ *are l. i. vectors of* $\mathbb{R}^n$*;*
(ii) $V_1, V_2, \dots, V_n$ *are l. i. vectors of* $C^1(I; \mathbb{R}^n)$*;*
(iii) *for every* $t \in I$, $V_1(t), V_2(t), \dots, V_n(t)$ *are l. i. vectors of* $\mathbb{R}^n$.

Proof. This is left as an exercise. □

Definition 1.3.2. A *fundamental matrix* of (1.3.5) is the $n \times n$ matrix formed by a fundamental system of solutions of (1.3.5).

Notation. Given an $n \times n$ matrix A, we denote by A^i its ith column and write

$$A = (A^1 | \dots | A^n).$$

With this notation, if we display a vector $X = (x_1, \dots, x_n) \in \mathbb{R}^n$ as a column vector as in (1.3.3), by the usual rules of product of matrices we see that

$$AX = x_1 A^1 + \cdots x_n A^n.$$

If $V_1, V_2, \dots, V_n$ are n linearly independent solutions of (1.3.5), let E denote the corresponding fundamental matrix:

$$E = (V_1 | \dots | V_n),$$

constructed putting $E^i = V_i$. With these notations, looking at formula (1.3.11) we see that the content of Theorem 1.3.1 can be resumed saying that any solution U of (1.3.5), i. e., any $U \in S_0$, can be written

$$U = EC \quad (C \in \mathbb{R}^n). \tag{1.3.12}$$

We now return to the inhomogeneous equation (1.3.4) and to its solution set S. Lemma 1.3.2, part (ii), tells us that it is enough to have *one* solution X_0 of (1.3.4) to have *all* solutions of it, because $S = S_0 + X_0$; and by what has just been said above about S_0, we conclude that the solutions of (1.3.4) are all given by the formula

$$U = EC + X_0 \quad (C \in \mathbb{R}^n).$$

To find an $X_0 \in S$, one uses **the method of variation of constants**, which consists in looking for an X_0 of the form

$$X_0(t) = E(t)C(t), \tag{1.3.13}$$

with E a fundamental matrix of (1.3.5); the terminology (due to Lagrange) explains that we are replacing in (1.3.12) the constant C with a *function* $C = C(t)$ to be determined. To do this, impose that EC be a solution of (1.3.4); this means that

$$\frac{d}{dt}(E(t)C(t)) = A(t)E(t)C(t) + B(t) \quad (t \in I). \tag{1.3.14}$$

It is easy to check that

$$\frac{d}{dt}(E(t)C(t)) = E'(t)C(t) + E(t)C'(t)$$

and that

$$E'(t) = A(t)E(t)$$

(which shows that E is a *matrix solution* of (1.3.5)); using these two relations in (1.3.14) yields the following condition on C:

$$E(t)C'(t) = B(t).$$

Thanks to the invertibility of E, this is equivalent to the condition $C' = E^{-1}B$ and can be satisfied taking for C any primitive of the (continuous) vector function $E^{-1}B$.

We can finally resume the information about the linear systems (1.3.4) in the following theorem.

Theorem 1.3.2. *The solutions of the linear system* (1.3.4) *are all given by the formula*

$$U(t) = E(t)C + E(t)\int [E(t)]^{-1}B(t)\,dt, \tag{1.3.15}$$

where E is a fundamental matrix of the homogeneous system (1.3.5), $C \in \mathbb{R}^n$, *and for $F \in C(I, \mathbb{R}^n)$, the symbol $\int F(t)\,dt$ denotes an arbitrarily chosen primitive of F in I.*

Remark 1.3.2. The reader is invited to appreciate both the beauty of formula (1.3.15) and its strength as an extension of formula (1.1.2) with which this chapter has begun.

1.4 Linear systems with constant coefficients. The exponential matrix

Recall that for every $x \in \mathbb{R}$,

$$e^x = \sum_{n=0}^{\infty} \frac{x^n}{n!} = 1 + x + \frac{x^2}{2!} + \cdots. \tag{1.4.1}$$

Definition 1.4.1. Given any $n \times n$ matrix A, define

$$e^A = \sum_{n=0}^{\infty} \frac{A^n}{n!} = I + A + \frac{A^2}{2!} + \cdots, \tag{1.4.2}$$

where $A^n = A \times \cdots \times A$ is the n-times usual product row by columns of A times itself.

In this section, we show that e^{tA} is a fundamental matrix for the system of ODEs

$$X' = AX \tag{1.4.3}$$

so that, on the basis of Theorem 1.3.2, the solutions of the system

$$X' = AX + B(t) \tag{1.4.4}$$

will be given by the formula

$$U(t) = e^{tA}C + e^{tA}\int e^{-tA}B(t)\,dt \quad (C \in \mathbb{R}^n). \tag{1.4.5}$$

Recall that in explicit form, if $A = (a_{ij})$, $1 \le i,j \le n$, (1.4.4) is written

$$\begin{cases} x_1' = a_{11}x_1 + a_{12}x_2 + \cdots + a_{1n}x_n + b_1(t) \\ x_2' = a_{21}x_1 + a_{22}x_2 + \cdots + a_{2n}x_n + b_2(t) \\ \dots \\ x_n' = a_{n1}x_1 + a_{n2}x_2 + \cdots + a_{nn}x_n + b_n(t). \end{cases} \tag{1.4.6}$$

Two points have to be discussed:
- the actual convergence of the series of matrices (1.4.2);
- the verification of e^{tA} being a fundamental matrix for (1.4.3).

A. Series of matrices and their convergence

Let

$$M_n \equiv \{A = (a_{ij}),\ 1 \le i,j \le n\} \tag{1.4.7}$$

denote the set of all $n \times n$ real matrices. M_n becomes a vector space when equipped with the usual operations of sum and product by a real number. A **norm** in this vector space can be introduced in a natural way putting

$$\|A\| = \sqrt{\sum_{i,j=1}^{n} a_{i,j}^2}. \tag{1.4.8}$$

A very useful criterion for the convergence of series in a **Banach** space, which is a **complete** normed vector space (see Section 2.4 of Chapter 2), is given by the following theorem, which extends the well-known criterion of the **absolute convergence** for a numerical series.

Theorem 1.4.1. *Let E be a Banach space and let (x_n) be a sequence in E. If the numerical series $\sum_{n=1}^{\infty} \|x_n\|$ converges, then the series $\sum_{n=1}^{\infty} x_n$ converges in E.*

The proof of Theorem 1.4.1 is essentially the same as that for numerical series, and can be seen in Chapter 2, Section 2.4.

Consider the particular space M_n of the real matrices, normed via (1.4.8). First note that this is a Banach space as it is isometrically isomorphic to $\mathbb{R}^{n^2}$. Moreover, here we have an additional and useful property of the norm.

Lemma 1.4.1. *Let $A, B \in M_n$. Then*

$$\|AB\| \leq \|A\|\|B\|. \tag{1.4.9}$$

Proof. This is left as an exercise (recall that $(AB)_{ij} = A_i \cdot B^j$ and use Schwarz' inequality). □

Proposition 1.4.1. *For any $A \in M_n$, the series*

$$\sum_{n=1}^{\infty} \frac{A^n}{n!} = I + A + \frac{A^2}{2!} + \cdots$$

converges in the normed space M_n. Its sum is denoted with e^A and is called the exponential matrix *of A.*

Proof. On the basis of Theorem 1.4.1 and the completeness of M_n, it is enough to show that the series

$$\sum_{n=0}^{\infty} \frac{\|A^n\|}{n!} = 1 + \|A\| + \cdots + \frac{\|A^n\|}{n!} + \cdots \tag{1.4.10}$$

converges. However, by virtue of Lemma 1.4.1 we have

$$\|A^n\| \leq \|A\|^n \quad (n \in \mathbb{N})$$

so that the series (1.4.10) is term-by-term majorized by the convergent series $\sum_{n=0}^{\infty} \|A\|^n/n!$ (whose sum is $e^{\|A\|}$, see (1.4.1)), and thus converges by the familiar comparison criterion for series with non-negative entries. □

B. The exponential matrix e^{tA} as a fundamental matrix for $X' = AX$

Let us first remark some properties of the **matrices of solutions** of a general linear homogeneous system

$$X' = A(t)X. \tag{1.4.11}$$

Proposition 1.4.2. *Let $V_1, V_2, \ldots, V_n \in C^1(I, \mathbb{R}^n)$ and let E denote the corresponding matrix*

$$E = (V_1|\ldots|V_n)$$

constructed putting $E^i = V_i$ $(1 \le i \le n)$*. Then the following properties are equivalent:*
(i) $V_1, V_2, \dots, V_n$ *are solutions of* (1.4.11);
(ii) E *is a* ***matrix solution*** *of* (1.4.11), *that is,*

$$E'(t) = A(t)E(t) \quad \forall t \in I; \tag{1.4.12}$$

(iii) *for every* $C \in \mathbb{R}^n$, $X(t) = E(t)C$ *is a solution of* (1.4.11).

Proof. (i) $\Rightarrow$ (ii) (already seen in the method of variation of constants): The matrix equality in (1.4.12) can be verified "by columns" and – since $(AB)^i = AB^i$ for any $A, B \in M_n$ – is thus equivalent to statement (i).

(ii) $\Rightarrow$ (iii): If (1.4.12) holds, then

$$\frac{d}{dt}(E(t)C) = E'(t)C = A(t)E(t)C \quad \forall t \in I.$$

(iii) $\Rightarrow$ (i): By assumption, $E(t)C = c_1V_1 + \cdots c_nV_n$ is a solution of (1.4.11) for every $C \in \mathbb{R}^n$. Thus, taking in particular $C = \mathbf{e}_i$, it follows that V_i is a solution for every i. □

Proposition 1.4.3. *If* $E = E(t)$ *is a matrix solution of* (1.4.11), *then* E *is fundamental* $\iff$ $E(t)$ *is invertible* $\forall t \in I \iff \exists \hat{t} \in I : E(\hat{t})$ *is invertible.*

Proof. This follows from Proposition 1.3.3, which gives equivalent criteria for the linear independence of n solutions of (1.4.11). □

Let us go back to the exponential matrix. By Proposition 1.4.1, we can consider for every $t \in \mathbb{R}$ the matrix

$$e^{tA} = \sum_{n=0}^{\infty} \frac{(tA)^n}{n!} = I + tA + \frac{t^2A^2}{2!} + \cdots + \frac{t^kA^k}{k!} + \cdots. \tag{1.4.13}$$

Theorem 1.4.2. *Given the linear system with constant coefficients* $X' = AX$, *the matrix* $E(t) = e^{tA}$ *is a fundamental matrix for it. Therefore, the general solution of* $X' = AX$ *is*

$$X(t) = e^{tA}C, \quad C \in \mathbb{R}^n. \tag{1.4.14}$$

Proof. Due to Propositions 1.4.2 and 1.4.3, it will be enough to show that
- $\frac{d}{dt}(e^{tA}) = Ae^{tA}$ $\forall t \in \mathbb{R}$;
- $e^{tA}|_{t=0} = I$.

While the second equality is an immediate consequence of the definition (1.4.13), the first requires some more care. For every i, j with $1 \le i, j \le n$ we have

$$\left[\frac{d}{dt}(e^{tA})\right]_{ij} = \frac{d}{dt}(e^{tA})_{ij} = \frac{d}{dt}\left(\sum_{n=0}^{\infty}\frac{(tA)^n}{n!}\right)_{ij} = \frac{d}{dt}\left(\sum_{n=0}^{\infty}\frac{t^n}{n!}(A^n)_{ij}\right)$$

$$= \sum_{n=0}^{\infty} \frac{d}{dt}\left(\frac{t^n}{n!}(A^n)_{ij}\right) = \sum_{n=1}^{\infty} n\frac{t^{n-1}}{n!}(A^n)_{ij} = \sum_{k=0}^{\infty} \frac{t^k}{k!}(A^{k+1})_{ij}$$
$$= \left(\sum_{k=0}^{\infty} \frac{t^k}{k!}(A^{k+1})\right)_{ij},$$

the equality sign between the first and the second line being allowed because power series can be differentiated term-by-term, as will be recalled in Section 2.7 of Chapter 2. Therefore,

$$\frac{d}{dt}(e^{tA}) = \sum_{k=0}^{\infty} \frac{t^k}{k!}A^{k+1} = A\sum_{k=0}^{\infty} \frac{t^k}{k!}A^k = Ae^{tA}. \qquad \square$$

Computation of e^{tA}

In order for formula (1.4.14) to be of practical use, methods for computing e^{tA} starting from a given A are necessary. The simplest case is when A is **diagonal**, for if $A = \operatorname{diag}(a_1, \ldots, a_n)$, then

$$A^k = \begin{pmatrix} a_1^k & \ldots & 0 \\ 0 & \ldots & 0 \\ \vdots & \ldots & \vdots \\ 0 & \ldots & a_n^k \end{pmatrix},$$

so

$$e^{tA} = \begin{pmatrix} e^{ta_1} & \ldots & 0 \\ 0 & \ldots & 0 \\ \vdots & \ldots & \vdots \\ 0 & \ldots & e^{ta_n} \end{pmatrix}, \tag{1.4.15}$$

because

$$(e^A)_{ij} = \sum_{k=0}^{\infty}\left(\frac{A^k}{k!}\right)_{ij} = \begin{cases} \sum_{k=0}^{\infty} \frac{a_i{}^k}{k!} = e^{a_i} & (i = j) \\ 0 & (i \neq j). \end{cases}$$

The next case is when A is **diagonable**, that is, when A is **similar** to a diagonal matrix, this in turn meaning that there is an invertible matrix P such that

$$P^{-1}AP = D, \tag{1.4.16}$$

with D diagonal. Indeed, the definition (1.4.2) shows that if $A, B \in M_n$ are similar, then so are their exponentials e^A, e^B, for if $B = P^{-1}AP$ for some invertible P, then

$$B^k = B \cdot B \cdot \cdots \cdot B = (P^{-1}AP) \cdot (P^{-1}AP) \cdot \cdots \cdot (P^{-1}AP) = P^{-1}A^kP,$$

so that using (1.4.2) we obtain $e^B = P^{-1}e^A P$. Therefore, (1.4.16) implies that $e^{tD} = P^{-1}e^{tA}P$, whence

$$e^{tA} = Pe^{tD}P^{-1}, \tag{1.4.17}$$

so that, utilizing (1.4.15) for e^{tD} and knowing the **similarity matrix** P, one can recover e^{tA}. However, knowing the similarity matrix P and the resulting diagonal matrix D amounts to knowing the **eigenvectors** of A and the corresponding **eigenvalues**. Indeed, provided that P is invertible we can write the equivalences

$$P^{-1}AP = D \Leftrightarrow AP = PD \Leftrightarrow (AP)^i = (PD)^i \Leftrightarrow AP^i = PD^i \quad (1 \le i \le n),$$

so that, putting $V_i = P^i$ and $D = \mathrm{diag}(\lambda_1, \dots, \lambda_n)$, we arrive at

$$AV_i = \lambda_i V_i \quad (1 \le i \le n). \tag{1.4.18}$$

The condition of invertibility on P is equivalent to the condition that V_i be linearly independent, that is, that they form a basis of $\mathbb{R}^n$ (assuming, as we are doing for simplicity, that A be diagonable in the real field, so that P and D are real matrices).

Theorem 1.4.3. *Suppose that $A \in M_n$ is diagonable in the real field and let $V_1, \dots, V_n$ be n linearly independent eigenvectors of A corresponding to the eigenvalues $\lambda_1, \dots, \lambda_n$ as in* (1.4.18). *Then the general solution of $X' = AX$ is*

$$X(t) = k_1 e^{\lambda_1 t} V_1 + \cdots k_n e^{\lambda_n t} V_n \quad (k_1, \dots, k_n \in \mathbb{R}). \tag{1.4.19}$$

In other words, the functions

$$X_1(t) = e^{\lambda_1 t} V_1, \dots, X_n(t) = e^{\lambda_n t} V_n$$

form a fundamental system of solutions of $X' = AX$.

Proof. From Theorem 1.4.2 and formula (1.4.17) we have, putting $K = P^{-1}C$,

$$X(t) = e^{tA}C = Pe^{tD}P^{-1}C = Pe^{tD}K. \tag{1.4.20}$$

However,

$$e^{tD} = \begin{pmatrix} e^{\lambda_1 t} & \dots & 0 \\ 0 & \dots & 0 \\ \vdots & \dots & \vdots \\ 0 & \dots & e^{\lambda_n t} \end{pmatrix}, \tag{1.4.21}$$

so that, with $K = (k_1, \dots, k_n)$,

$$e^{tD}K = k_1 e^{\lambda_1 t}\mathbf{e_1} + \cdots k_n e^{\lambda_n t}\mathbf{e_n}.$$

Putting this in (1.4.20) we obtain (1.4.19), since $P\mathbf{e}_i = P^i = V_i$. □

Remark 1.4.1. For the sake of simplicity, we will not deal with the important case in which the (real) coefficient matrix A has some complex (and necessarily conjugate) eigenvalues. Of course, A could still be diagonalized in the complex field, in which case Theorem 1.4.3 remains essentially unaltered: we refer for this to Hale [5] or Walter [1].

1.5 Higher-order ODEs

A. An *ordinary differential equation of order n* (briefly, an nth-order ODE) has the form

$$x^{(n)} = f(t, x, x', \dots, x^{(n-1)}), \tag{1.5.1}$$

where $f : A \to \mathbb{R}$ with $A \subset \mathbb{R}^{n+1}$. A **solution** of (1.5.1) is a function u such that

$$u^{(n)}(t) = f(t, u(t), u'(t), \dots, u^{(n-1)}(t)) \quad \forall t \in J_u, \tag{1.5.2}$$

where $J_u \subset \mathbb{R}$ is an interval. For this definition to make sense, it is of course required that u be n times differentiable in J_u and that the $(n+1)$-tuple of real numbers $(t, u(t), u'(t), \dots, u^{(n-1)}(t))$ belong to A for every $t \in J_u$.

Example 1.5.1. Consider the following:

$$x'' = 0, \quad u(t) = ct + d,$$

$$x'' = g = \text{const.}, \quad u(t) = \frac{1}{2}gt^2 + ct + d.$$

Example 1.5.2. More generally, given $g \in C(I)$,

$$x'' = g(t), \quad u(t) = \int_{t_0}^{t}\left(\int_{t_0}^{s} g(y)dy\right)ds + ct + d,$$

where t_0 is any point of I.

Example 1.5.3. Consider

$$mx'' = -kx, \quad u(t) = c\cos\omega t + d\sin\omega t, \quad \left(\omega = \sqrt{\frac{k}{m}}\right).$$

In the above examples, we see that the solution depends upon **two** arbitrary constants c and d, so that the solution will be uniquely determined if we assign **two** initial conditions ("initial position and velocity" in the mechanical interpretation).

The concepts of **maximal solution** and **global solution** (when $A = I \times \mathbb{R}^n$) of (1.5.1) are given in the same way as for the case $n = 1$.

Likewise, here too we observe the *regularization property* of solutions of an ODE: equality (1.5.2) shows that if f is **continuous**, then any solution u of (1.5.1) has a necessarily continuous nth derivative, so that $u \in C^n(J_u)$.

B. The IVP problem (Cauchy problem) for (1.5.1)

Definition 1.5.1. The *initial value problem (IVP)* for the nth-order ODE (1.5.1) consists – given an $(n+1)$-tuple of real numbers $(t_0, x_1^0, x_2^0, \dots, x_n^0) \in A$ – in finding a solution u of (1.5.1) such that:

(i) u is defined in a neighborhood of t_0;

(ii) $u(t_0) = x_1^0, u'(t_0) = x_2^0, \dots, u^{(n-1)}(t_0) = x_n^0$.

For instance, the IVP for a third-order ODE will be written as

$$\begin{cases} x''' = f(t, x, x', x'') \\ x(t_0) = x_0 \\ x'(t_0) = y_0 \\ x''(t_0) = z_0, \end{cases} \tag{1.5.3}$$

where (t_0, x_0, y_0, z_0) is a given point in $A \equiv \mathrm{dom}(f) \subset \mathbb{R}^4$.

Exercise 1.5.1. Check that the IVP

$$\begin{cases} x'' = t \\ x(t_0) = x_0 \\ x'(t_0) = y_0 \end{cases} \tag{1.5.4}$$

has a unique solution, which is given by the formula

$$u(t) = \frac{1}{6}(t^3 - t_0^3) - \frac{1}{2}t_0^2(t - t_0) + ty_0 + (x_0 - t_0 y_0).$$

Question. Do we have general existence and uniqueness results for solutions of the IVP for (1.5.1)? The next step will help to answer this question.

C. Equivalence between an *n*th-order ODE and a first-order system

Proposition 1.5.1. *The nth-order ODE* (1.5.1) *is equivalent to the system*

$$\begin{cases} x_1' = x_2 \\ x_2' = x_3 \\ \dots \\ x_n' = f(t, x_1, x_2, \dots, x_n) \end{cases} \tag{1.5.5}$$

in the following precise sense:

(a) *if* $u : J_u \to \mathbb{R}$ *is a solution of* (1.5.1), *then the vector function* $U \equiv (u, u', \ldots, u^{(n-1)})$ *is a solution of* (1.5.5);

(b) *vice versa, if* $U = (u_1, u_2, \ldots, u_n) : J_U \to \mathbb{R}^n$ *is a solution of* (1.5.5), *then its first component* u_1 *is a solution of* (1.5.1).

Proof (for the case $n = 2$*).* (a) Let $u : J_u \to \mathbb{R}$ be a solution of (1.5.1). Then, by definition,

$$u''(t) = f(t, u(t), u'(t)), \quad t \in J_u. \tag{1.5.6}$$

Consider the vector function $U = (u, u')$. Then U satisfies trivially the first equation of the system

$$\begin{cases} x_1' = x_2 \\ x_2' = f(t, x_1, x_2), \end{cases} \tag{1.5.7}$$

while the second one is satisfied by virtue of (1.5.6).

The verification of (b) is equally simple. □

D. Existence and uniqueness theorems for the IVP

Write (1.5.5) in the vector form $X' = F(t, X)$, where $F = (F_1, \ldots, F_n) : A \to \mathbb{R}^{n+1}$ is defined putting, for $X = (x_1, \ldots, x_n)$,

$$\begin{cases} F_i(t, X) = x_{i+1}, \quad 1 \le i \le n-1 \\ F_n(t, X) = f(t, x_1, \ldots, x_n). \end{cases} \tag{1.5.8}$$

Now observe that:
- F continuous $\Leftrightarrow f$ continuous,
- F locally Lipschitz $\Leftrightarrow f$ locally Lipschitz,
- F sublinear $\Leftrightarrow f$ sublinear,

because for every $1 \le i \le n-1$, F_i trivially satisfies these conditions.

Therefore, all existence and uniqueness theorems for the IVP for the system $X' = F(t, X)$ translate into existence and uniqueness theorems for the IVP for (1.5.1), under the corresponding assumptions (continuity, etc.) **upon** f.

For instance, for the case $n = 2$ we have the following theorem.

Theorem 1.5.1. *Consider the IVP*

$$\begin{cases} x'' = f(t, x, x') \\ x(t_0) = x_0, \quad x'(t_0) = y_0, \end{cases} \tag{1.5.9}$$

where $f = f(t, x, y)$ *is a real-valued function defined in an open subset* $A \subset \mathbb{R}^3$. *Suppose that* f *is:*

(a) *continuous in A;*
(b) *locally Lipschitz with respect to $X = (x,y)$ in A.*

Then for every $(t_0, x_0, y_0) \in A$, there exists an $r_0 > 0$ such that the IVP (1.5.9) *has a unique solution defined in the neighborhood* $]t_0 - r_0, t_0 + r_0[$ *of* t_0.

Proof. The equation $x'' = f(t,x,x')$ is equivalent to the system

$$\begin{cases} x' = y \\ y' = f(t,x,y). \end{cases} \tag{1.5.10}$$

As remarked before, our assumptions on f are inherited from the function $F : A \to \mathbb{R}^2$ defined putting

$$F(t,x,y) = (y, f(t,x,y))$$

and therefore, by virtue of the local existence and uniqueness theorem, Theorem 1.2.1, for first-order systems, the IVP

$$\begin{cases} X' = F(t,X) \\ X(t_0) = (x_0, y_0) \end{cases} \tag{1.5.11}$$

has a unique solution U defined in a neighborhood I_0 of t_0. By the special form of (1.5.10), we have

$$U = (u,v) = (u,u').$$

By Proposition 1.5.1, part (b), u is a solution of $x'' = f(t,x,x')$; moreover, since

$$U(t_0) = (u(t_0), u'(t_0) = (x_0, y_0)$$

we conclude that $u : I_0 \to \mathbb{R}$ solves our original problem (1.5.9). □

Example 1.5.4. Check the applicability of Theorem 1.5.1 for the equations

$$\text{(i) } x'' = x^2 + x'^2, \quad \text{(ii) } x'' = t\sin(xx'), \quad \text{(iii) } x'' = \frac{(\sin x)x'}{\sqrt{1-t^2}}.$$

Exercise 1.5.2.
- State and prove (following the same pattern shown in the proof of Theorem 1.5.1) an existence and uniqueness theorem for *global* solutions of (1.5.9).
- Where are the maximal solutions of equations (ii) and (iii) in Example 1.5.4 defined?

Exercise 1.5.3. Consider the equation

$$x'' = a(t)\sqrt{x^2 + x'^2}, \tag{1.5.12}$$

where $a \in C(I)$.

(a) Write the first-order system equivalent to (1.5.12).

(b) Show that the IVP for (1.5.12) – whatever the initial conditions – has a unique global solution.

(c) What is the solution u of (1.5.12) such that $u(1) = u'(1) = 0$?

1.6 Linear ODEs of higher order

A **linear** ordinary differential equation of order n is usually written as

$$x^{(n)} + a_1(t)x^{(n-1)} + \cdots + a_n(t)x = b(t), \tag{1.6.1}$$

where the coefficient functions a_i $(1 \le i \le n)$ and b are real-valued functions defined on an interval $I \subset \mathbb{R}$.

Example 1.6.1. Consider

$$x''' + (t \ln t)x' + \frac{x}{\sqrt{1-t^2}} = 1, \quad I =]0,1[.$$

The first-order system equivalent to (1.6.1) is

$$\begin{cases} x_1' = x_2 \\ x_2' = x_3 \\ \cdots \\ x_{n-1}' = x_n \\ x_n' = -a_1(t)x_n + \cdots - a_n(t)x_1 + b(t) \equiv f(t, x_1, x_2, \ldots, x_n), \end{cases} \tag{1.6.2}$$

with $f : I \times \mathbb{R}^n \to \mathbb{R}$. As usual, we write the system (1.6.2) in vector form:

$$X' = A(t)X + B(t),$$

with

$$A = \begin{pmatrix} 0 & 1 & 0 & \cdots & 0 \\ 0 & 0 & 1 & \cdots & 0 \\ \vdots & \vdots & \vdots & \cdots & \vdots \\ 0 & 0 & 0 & \cdots & 1 \\ -a_n & -a_{n-1} & \cdots & -a_2 & -a_1 \end{pmatrix}, \quad B = \begin{pmatrix} 0 \\ 0 \\ \cdots \\ 0 \\ b \end{pmatrix}.$$

Thus, we shall use (i) the results about the **equivalence** between an nth-order ODE and a first-order system and (ii) the results about **linear** systems.

Proposition 1.6.1. *Let*

$$A = \{u : u \text{ is a solution of } (1.5.1)\},$$
$$\hat{A} = \{U : U \text{ is a solution of } (1.5.5)\}$$

and let $G : A \to \hat{A}$, $H : \hat{A} \to A$ be the maps defined as follows:

$$G(u) = (u, u', \dots, u^{(n-1)}) \quad (u \in A), \tag{1.6.3}$$
$$H(U) = u_1 \quad (U = (u_1, \dots, u_n) \in \hat{A}). \tag{1.6.4}$$

Then

$$H(G(u)) = u \quad \forall u \in A \quad \text{and} \quad G(H(U)) = U \quad \forall U \in \hat{A}. \tag{1.6.5}$$

In other words, G is a ***bijective*** *map of A onto $\hat{A}$, whose inverse map is H.*

Proof. Use Proposition 1.5.1 and check the equalities in (1.6.5), using in particular the fact that, due to the special form of the system (1.5.5), we have

$$U = (u_1, u_2, \dots, u_n) = (u_1, u_1', \dots, u_1^{(n-1)})$$

for any $U \in \hat{A}$.

From now on, we shall always suppose that the coefficients a_i and b of (1.6.1) are **continuous** on I. It then follows that any (maximal) solution u of (1.6.1) is (i) defined on I and (ii) of class C^n; that is, $u \in C^n(I)$.

As done with first-order linear equations and systems, we consider the homogeneous equation associated with (1.6.1):

$$x^{(n)} + a_1(t)x^{n-1} + \dots + a_n(t)x = 0, \tag{1.6.6}$$

which is equivalent to the homogeneous linear system

$$X' = A(t)X. \tag{1.6.7}$$

□

Theorem 1.6.1. *The set of all solutions of (1.6.6) is a vector space of dimension n. More precisely, it is an n-dimensional vector subspace of $C^n(I)$.*

Proof. (a) Let S_0 and $\hat{S}_0$ be the sets of solutions of (1.6.6) and (1.6.7), respectively. They are both vector spaces (precisely, subspaces of $C^n(I)$ and $C^1(I, \mathbb{R}^n)$, respectively) by the linearity and homogeneity of the equations involved.

(b) The map $G : S_0 \to \hat{S}_0$ defined in Proposition 1.6.1 putting

$$G(u) = (u, u', \dots, u^{(n-1)}) \quad (u \in S_0)$$

is evidently linear, and by the same proposition is thus an **isomorphism** of S_0 onto $\hat{S}_0$. Since $\dim \hat{S}_0 = n$ (by Theorem 1.3.1), we conclude that also $\dim S_0 = n$. □

General solution of (1.6.1) and its relation with the general solution of the homogeneous equation

It follows from Theorem 1.6.1 that if $v_1, \dots v_n$ are n linearly independent solutions of (1.6.6), then any solution u of this equation is a linear combination of them:

$$u = c_1 v_1 + \cdots c_n v_n.$$

As to the non-homogeneous equation (1.6.1), by the same argument used for linear systems (that is, on the basis of Proposition 1.3.1), we have the following corollary.

Corollary 1.6.1. *If $v_1, \dots v_n$ are linearly independent solutions of* (1.6.6) *and z is a solution of* (1.6.1), *then any solution u of* (1.6.1) *can be written as*

$$u = c_1 v_1 + \cdots c_n v_n + z,$$

with $c_1, \dots, c_n \in \mathbb{R}$.

Criterion for the linear independence of solutions: Wronskian determinant

Proposition 1.6.2. *Let $v_1, \dots v_n \in C^n(I)$ be n solutions of the homogeneous equation* (1.6.6) *and let W be their* Wronskian determinant, *defined putting*

$$W(t) = \det \begin{pmatrix} v_1(t) & \dots & v_n(t) \\ v_1'(t) & \dots & v_n'(t) \\ \vdots & \dots & \vdots \\ v_1^{(n-1)}(t) & \dots & v_n^{(n-1)(t)} \end{pmatrix}, \quad t \in I. \tag{1.6.8}$$

Then $v_1, \dots v_n$ are ***linearly independent*** *if and only if, for any $t \in I$, $W(t) \neq 0$.*

Proof. Given $v_1, \dots v_n$, put

$$V_i = (v_i, \dots v_i^{(n-1)}), \quad 1 \le i \le n-1,$$

so that V_i is the ith column of the n-by-n matrix in (1.6.8). From the properties of the map G defined in (1.6.3), we know that

$v_1, \dots v_n$ linearly independent solutions of (1.6.6) $\Leftrightarrow$
$V_1 = G(v_1), \dots, V_n = G(v_n)$ linearly independent solutions of (1.6.7).

Moreover, from Proposition 1.3.3, we know that

$V_1, \dots, V_n$ linearly independent solutions of (1.6.7) $\Leftrightarrow$
$\forall t \in I, V_1(t), \dots, V_n(t)$ linearly independent vectors of $\mathbb{R}^n$,

so that the conclusion of Proposition 1.6.2 follows from the familiar criterion of l. i. of vectors in $\mathbb{R}^n$ via determinants, since

$$W(t) = \det(V_1(t)|\dots|V_n(t)).$$

Moreover, we have (again by Proposition 1.3.3)

$$W(t) \neq 0 \quad \forall t \in I \Leftrightarrow \exists \hat{t} \in I : W(\hat{t}) \neq 0.$$ □

Example 1.6.2. Consider

$$x'' - \frac{2x}{t^2} = t, \quad t \in]0, +\infty[\equiv I. \tag{1.6.9}$$

If we consider the homogeneous equation associated with (1.6.9) and look for solutions of the form t^α, we easily find the two solutions

$$v_1(t) = t^2, \quad v_2(t) = \frac{1}{t},$$

which are linearly independent, because for all t

$$W(t) = -3.$$

On the basis of Corollary 1.6.1, in order to find all solutions of (1.6.9) it remains to find one particular solution of the same equation. We do this using the method of variation of constants, discussed for linear systems in Section 1.3.

Let us see here how **the method of variation of constants** works for second-order linear ODEs

$$x'' + a_1(t)x' + a_2(t)x = b(t). \tag{1.6.10}$$

Let v_1, v_2 be two linearly independent solutions of the homogeneous equation

$$x'' + a_1(t)x' + a_2(t)x = 0. \tag{1.6.11}$$

The system of two first-order equations equivalent to (1.6.10) is (see (1.6.2))

$$\begin{cases} x' = y \\ y' = -a_2(t)x - a_1(t)y + b(t). \end{cases} \tag{1.6.12}$$

We have

$$v_1, v_2 \quad \text{l. i. solutions of (1.6.11)} \Leftrightarrow V_1 \equiv \begin{pmatrix} v_1 \\ v_1' \end{pmatrix}, V_2 \equiv \begin{pmatrix} v_1 \\ v_1' \end{pmatrix} \quad \text{l. i. solutions of}$$

the homogeneous system corresponding to (1.6.12). Now recall from the section on linear systems (Section 1.3) that a solution X_0 of (1.6.12) is obtained by the formula

$$X_0(t) = E(t) \int [E(t)]^{-1} B(t)\, dt,$$

where E is a fundamental matrix. In our case, we take

$$E = (V_1 | V_2) = \begin{pmatrix} v_1 & v_2 \\ v_1' & v_2' \end{pmatrix}.$$

Recall that if

$$A = \begin{pmatrix} a & b \\ c & d \end{pmatrix}$$

with $\det A \neq 0$, then the inverse matrix A^{-1} is given by

$$A^{-1} = \frac{1}{\det A} \begin{pmatrix} d & -b \\ -c & a \end{pmatrix}.$$

Therefore,

$$\begin{aligned} E^{-1} &= \frac{1}{W} \begin{pmatrix} v_2' & -v_2 \\ -v_1' & v_1 \end{pmatrix}, \\ E^{-1}B &= \frac{1}{W} \begin{pmatrix} v_2' & -v_2 \\ -v_1' & v_1 \end{pmatrix} \begin{pmatrix} 0 \\ b \end{pmatrix} = \frac{1}{W} \begin{pmatrix} -v_2 b \\ v_1 b \end{pmatrix}, \\ X_0 &= E \int E^{-1} B = \begin{pmatrix} v_1 & v_2 \\ v_1' & v_2' \end{pmatrix} \int \frac{1}{W} \begin{pmatrix} -v_2 b \\ v_1 b \end{pmatrix} = \\ &= \begin{pmatrix} v_1 \int \frac{1}{W}(-v_2 b) + v_2 \int \frac{1}{W}(v_1 b) \\ \cdots \end{pmatrix}. \end{aligned}$$

The first component z of X_0 is a solution of (1.6.10):

$$z(t) = v_1(t) \int \frac{-v_2(t) b(t)}{W(t)}\, dt + v_2(t) \int \frac{v_1(t) b(t)}{W(t)}\, dt. \tag{1.6.13}$$

Using formula (1.6.13) in Example 1.6.2, we easily find that $z(t) = t^3/4$, so that the general solution of (1.6.9) is given by

$$u(t) = ct^2 + \frac{d}{t} + \frac{t^3}{4} \quad (c, d \in \mathbb{R}).$$

We give the following **exercises** on second-order linear ODEs:

$$x'' + a_1(t) x' + a_2(t) x = b(t), \tag{1.6.14}$$

$$x'' + a_1(t)x' + a_2(t)x = 0. \tag{1.6.15}$$

We have seen that if v_1, v_2 are l. i. solutions of (1.6.15), then the general solution of (1.6.14) is

$$u(t) = cv_1(t) + dv_2(t) + z(t),$$

where $c, d \in \mathbb{R}$ and z is given by (1.6.13).

Consider the special case in which (1.6.15) has **constant coefficients**:

$$x'' + a_1x' + a_2x = 0. \tag{1.6.16}$$

Then there is a simple rule (that can be recovered for instance by the results of Section 1.4 on systems) to find two independent solutions v_1, v_2: consider the **characteristic equation** associated with (1.6.16),

$$\lambda^2 + a_1\lambda + a_2 = 0,$$

and let λ_1, λ_2 be its roots. Then if these are real and distinct, take

$$v_1(t) = e^{\lambda_1 t}, \quad v_2(t) = e^{\lambda_2 t}.$$

If they are complex conjugate, so that $\lambda_{1,2} = \alpha \pm i\beta$ ($\beta \neq 0$), take

$$v_1(t) = e^{\alpha t}\cos\beta t, \quad v_2(t) = e^{\alpha t}\sin\beta t.$$

Finally, if $\lambda_1 = \lambda_2 \equiv \lambda$, take

$$v_1(t) = e^{\lambda t}, \quad v_2(t) = te^{\lambda t}.$$

Using these simple rules and (1.6.13), solve the following exercises.

Exercise 1.6.1. Check that the general solution of the equation

$$x'' - x = \frac{1}{1+e^t} \tag{1.6.17}$$

is

$$u(t) = ce^t + de^{-t} + \frac{e^t}{2}[-t - e^{-t} + \ln(1+e^t)] - \frac{e^{-t}}{2}\ln(1+e^t).$$

Exercise 1.6.2. Write the general solution of the following equations:

$$x'' - 3x' + 2x = t, \tag{1.6.18}$$

$$x'' + 4x = \tan t \quad \left(0 < t < \frac{\pi}{2}\right). \tag{1.6.19}$$

1.7 Additions and exercises

A1. Boundary value problems for linear second-order ODEs

In this chapter, we have taken as a starting point for our study of ODEs the *Cauchy problem* (or *initial value problem* [IVP]) – that is, the question of the existence of a solution to the differential equation that satisfies additional condition(s) in a given point t_0 belonging to the domain of the independent variable t. For this kind of problem, we have recalled or established existence and uniqueness results for the solution and studied the implications of this in particular to *linear* equations and systems; for instance, the "dimensional theorem," Theorem 1.3.1, is a nearly immediate, but nonetheless fundamental, consequence of these results.

If we look in particular at second-order linear equations, the IVP for them has the form

$$\begin{cases} x'' + a_1(t)x' + a_2(t)x = y(t), & t \in I \\ x(t_0) = \alpha \\ x'(t_0) = \beta, \end{cases} \tag{1.7.1}$$

where I is the interval on which the coefficients a_1, a_2 are defined and continuous and $t_0 \in I$. By the arguments discussed in Sections 1.5 and 1.6 (see in particular Exercise 1.5.2 applied to linear equations) we conclude that (1.7.1) has – for each given $\alpha, \beta \in \mathbb{R}$ and $y \in C(I)$ – a unique solution that is defined on the whole of I.

The situation just described may change quite dramatically if we add to the equation different kinds of supplementary conditions to be satisfied by the solution, as the following very simple examples show.

Example 1.7.1. The problem

$$\begin{cases} x'' = k \ (= \text{const.}) \\ x(0) = \alpha \\ x(1) = \beta \end{cases}$$

has a unique solution whatever α and β; it is given by

$$x(t) = \left(\beta - \alpha - \frac{k}{2}\right)t + \alpha + \frac{k}{2}t^2.$$

Example 1.7.2. The problem

$$\begin{cases} x'' + x = 0 \\ x(0) = 0 \\ x(\pi) = 0 \end{cases}$$

has infinitely many solutions:

$$x(t) = C \sin t, \quad C \in \mathbb{R}.$$

Example 1.7.3. The problem

$$\begin{cases} x'' + x = \sin t \\ x(0) = 0 \\ x(\pi) = 0 \end{cases}$$

has **no** solution whatsoever. Indeed, if there were such a solution x, then multiplying both members of the equation (which has become an equality!) by $\sin t$ and integrating we would have

$$\int_0^\pi [x''(t) + x(t)] \sin t dt = \int_0^\pi \sin^2 t dt = \frac{\pi}{2}. \tag{1.7.2}$$

However, a repeated integration by parts yields

$$\begin{aligned} \int_0^\pi x''(t) \sin t dt &= [x'(t) \sin t]_0^\pi - [x(t) \cos t]_0^\pi - \int_0^\pi x(t) \sin t dt \\ &= -\int_0^\pi x(t) \sin t dt. \end{aligned}$$

Hence the left-hand side in equality (1.7.2) is 0, contradicting the equality itself.

In order to understand this variety of situations, we simply have to go back to what we have seen in Section 1.6 about the general solution of linear second-order equations. For the sequel of this section we put

$$Lx = x'' + a_1(t)x' + a_2(t)x, \tag{1.7.3}$$

where $a_1, a_2 \in C([a, b])$, and consider the problems

$$(\mathbf{BVP}) \begin{cases} Lx = y(t), & a < t < b \\ x(a) = \alpha \\ x(b) = \beta, \end{cases} \qquad (\mathbf{BVP0}) \begin{cases} Lx = 0, & a < t < b \\ x(a) = 0 \\ x(b) = 0. \end{cases}$$

Theorem 1.7.1. *The problem* (**BVP**) *has a solution for any given* $y \in C([a, b])$ *and any given* $\alpha, \beta \in \mathbb{R}$ ***if and only if*** *the homogeneous problem* (**BVP0**) *has only the trivial solution* $u \equiv 0$. *Moreover, in this case, the solution to* (**BVP**) *is* ***unique***.

Proof. As we have seen in Corollary 1.6.1, the solutions of the second-order linear differential equation $Lx = y(t)$ are all given by the formula

$$u(t) = cv_1(t) + dv_2(t) + z(t) \quad c, d \in \mathbb{R}, \tag{1.7.4}$$

where v_1, v_2 are any two linearly independent solutions of the homogeneous equation $Lx = 0$ and z is any particular solution of $Lx = y$. So the question is: in equation (1.7.4), can we find $c, d \in \mathbb{R}$ so that u satisfies also the boundary conditions in (**BVP**)? This is quite easy to answer, for imposing the boundary conditions on u we find the linear algebraic system

$$\begin{cases} cv_1(a) + dv_2(a) = \alpha - z(a) \\ cv_1(b) + dv_2(b) = \beta - z(b) \end{cases}$$

and this will have a (unique) solution (c, d) for any given α, β and z if and only if the homogeneous system

$$\begin{cases} cv_1(a) + dv_2(a) = 0 \\ cv_1(b) + dv_2(b) = 0 \end{cases}$$

has *only* the solution $c = d = 0$. However, this is precisely the case in which (**BVP0**) has only the trivial solution $u = 0$. □

Remark 1.7.1. The proof of Theorem 1.7.1 shows that using formula (1.7.4), representing the "general solution" of a second-order linear differential equation, the existence (and uniqueness) problem for (**BVP**) reduces to recalling the basic property of linear maps of $\mathbb{R}^n$ into itself of being **surjective** if and only if they are **injective**.

Remark 1.7.2. The proof also shows that it works for more general BVPs such as

$$(\mathbf{BVP}) \begin{cases} Lx = y(t) & a < t < b \\ B_a[x] = \alpha \\ B_b[x] = \beta, \end{cases}$$

where B_a, B_b are linear boundary operators acting on x of the form

$$B_a(x) = px(a) + qx'(a) \quad (p, q \in \mathbb{R})$$

and similarly for B_b.

Exercise 1.7.1. Interpret Examples 1.7.1 to 1.7.3 in the light of Theorem 1.7.1.

Exercises

E1. Solutions of some of the exercises given in the text

Section 1.1

Exercise 1.1

By Definition 1.1.1, if we say $u : J_u \to \mathbb{R}$ is a solution of the differential equation $x' = f(t, x)$, this means that u is differentiable in J_u and that $u' = z_u$, with $z_u(t) = f(t, u(t))$ for $t \in J_u$.

Now the map z_u of J_u into $\mathbb{R}$ can be written as the composition $z_u = f \circ H$, where $H : J_u \to \mathbb{R}^2$ is defined via

$$H(t) = (t, u(t)), \quad t \in J_u.$$

Note that the composition is well defined because, by proviso (b) of Definition 1.1.1, $H(t) \in A$ (the domain of f) for every $t \in J_u$.

As u (being differentiable) is continuous, it follows that H is continuous; and since we know that the composition of continuous functions is a continuous function (as recalled in Theorem 0.0.1 of the Preliminaries), the continuity of z_u – that is, of u' – follows from that of f. Therefore, in this case u is not only differentiable, but of class C^1 in J_u.

Section 1.2

Exercise 2.1

The domain of definition of the given F,

$$F(t,x,y) = \left(tx \sin y + 1, \frac{\sqrt{x^2+y^2}}{t} \right) \equiv (f(t,x,y), g(t,x,y)),$$

is the open set $A = (\mathbb{R} \setminus \{0\}) \times \mathbb{R}^2$. Moreover, f and g are continuous in A and have continuous partial derivatives in the open set

$$A_0 = (\mathbb{R} \setminus \{0\}) \times (\mathbb{R}^2 \setminus \{(0,0)\}).$$

This guarantees the existence and uniqueness of a local solution to the differential system for any initial condition

$$x(t_0) = x_0, y(t_0) = y_0 \quad \text{with } t_0 \neq 0 \quad \text{and} \quad (x_0, y_0) \neq (0,0).$$

However, this does not prove the statement contained in the text of Exercise 1.2.1. To this purpose, since $t_0 = 1$ in equation (1.2.11), it is useful to consider the strip

$$S =]0, +\infty[\times \mathbb{R}^2 \subset A.$$

As noted in the text of Exercise 1.2.1, the lack of partial derivatives of g in S is overcome by the fact that g (and therefore F) is locally Lipschitz continuous with respect to $X = (x,y)$ in S; indeed,

$$|g(t,X) - g(t,Y)| \leq \frac{1}{|t|}\|X - Y\| \quad \forall t \neq 0, \quad \forall X, Y \in \mathbb{R}^2,$$

so that

$$|g(t,X) - g(t,Y)| \leq K\|X - Y\|$$

as long as t varies in a small neighborhood of a point $t_0 \neq 0$.

Finally, f and g are both sublinear with respect to X in S, for we have

$$|f(t,X)| = |tx \sin y + 1| \leq |t||x| + 1 \leq |t|\|X\| + 1 \equiv \alpha(t)\|X\| + \beta(t)$$

and

$$|g(t,X)| = \frac{\|X\|}{|t|} \equiv \gamma(t)\|X\|.$$

The desired statement is now proved by applying the global existence theorem, Theorem 1.2.2.

Section 1.3

Proof of Proposition 1.3.1

In the statement of Proposition 1.3.1 and related results, we use the notation $A + B$ (with A, B subsets of a given vector space E) in the obvious way, that is,

$$A + B \equiv \{a + b \mid a \in A, b \in B\}.$$

In case B is the singleton $\{c\}$, we write for convenience $A + c$ rather than $A + \{c\}$.

(i) Let us first prove that $S_y \subset \ker T + x_0$, with x_0 a fixed element in S_y. Pick $x \in S_y$ (thus by definition of S_y, $Tx = Tx_0 = y$) and write $x = x - x_0 + x_0 \equiv z + x_0$. Now $z \in \ker T$, for

$$Tz = T(x - x_0) = Tx - Tx_0 = y - y = 0.$$

This shows that $x = z + x_0 \in \ker T + x_0$.

(ii) Vice versa, to show that $\ker T + x_0 \subset S_y$, let $x \in \ker T + x_0$, so that $x = z + x_0$ for some $z \in \ker T$. Then

$$Tx = Tz + Tx_0 = y$$

so that $x \in S_y$.

Proof of Proposition 1.3.2

Let the vectors $x_1, \ldots, x_n$ form a basis of E; we claim that if $T : E \to F$ is an isomorphism, then the vectors $y_i \equiv Tx_i \quad (i = 1, \ldots, n)$ form a basis of F.

(a) $y_1, \ldots, y_n$ are linearly independent vectors of F, for if $c_1 y_1 + \cdots c_n y_n = 0$ for some $c_1, \ldots, c_n \in \mathbb{R}$, then

$$0 = c_1 Tx_1 + \cdots c_n Tx_n = T(c_1 x_1 + \cdots c_n x_n) \equiv Tz.$$

As T is injective, this implies that $z = 0$, and the l. i. of $x_1, \dots, x_n$ then implies that $c_1 = \dots = c_n = 0$.

(b) $y_1, \dots, y_n$ also generate F. Indeed, let $y \in F$ and (by the surjectivity of T) let $x \in E$ be such that $y = Tx$. Write $x = c_1x_1 + \cdots c_nx_n$ for suitable $c_1, \dots, c_n$ and see that

$$y = T(c_1x_1 + \cdots c_nx_n) = c_1y_1 + \cdots c_ny_n,$$

showing that y is a linear combination of $y_1, \dots, y_n$.

Proof of Proposition 1.3.3

(i) $\Rightarrow$ (ii): Suppose that $V_1(\hat{t}), V_2(\hat{t}), \dots, V_n(\hat{t})$ are l. i. and let $c_1, \dots, c_n \in \mathbb{R}$ be such that

$$c_1V_1 + \cdots c_nV_n = 0. \tag{1.7.5}$$

This is an equality between vectors of $C^1(I; \mathbb{R}^n)$, which are functions defined on I, and thus (1.7.5) means precisely that

$$c_1V_1(t) + \cdots c_nV_n(t) = 0 \quad \forall t \in I. \tag{1.7.6}$$

Note that the same symbol 0 denotes in (1.7.5) the zero of the vector space $C^1(I; \mathbb{R}^n)$ and in (1.7.6) the zero of the vector space $\mathbb{R}^n$. Taking $t = \hat{t}$ in (1.7.6), we conclude from our assumption that $c_1 = \dots = c_n = 0$, thus proving the l. i. of the vectors V_i.

(ii) $\Rightarrow$ (iii): Suppose now that $V_1, V_2, \dots, V_n$ are l. i. vectors of $C^1(I; \mathbb{R}^n)$. Take any $t_0 \in I$ and let $c_1, \dots, c_n \in \mathbb{R}$ be such that

$$c_1V_1(t_0) + \cdots c_nV_n(t_0) = 0. \tag{1.7.7}$$

We want to show that $c_i = 0$ for all i. Indeed, consider the function

$$Z \equiv c_1V_1 + \cdots c_nV_n,$$

and consider that (a) Z is a solution of (1.3.5) and (b) $Z(t_0) = 0$. Thus, by the uniqueness theorem, Z must be the identically zero solution; that is, $Z = 0$ in $C^1(I; \mathbb{R}^n)$. The definition of Z and our assumption now imply that $c_i = 0$ for all i.

The implication (iii) $\Rightarrow$ (i) is obvious.

Section 1.5

Exercise 5.2

1. Assuming that the domain A of the definition of f is a "solid strip" $I \times \mathbb{R}^2$ in $\mathbb{R}^3$, a condition to be added to assumptions (a) and (b) of Theorem 1.5.1 in order to have solutions of $x'' = f(t, x, x')$ defined on I is that f be sublinear in $X = (x, y)$, meaning that it satisfies in the strip an inequality of the form

$$|f(t,X)| = |f(t,x,y)| \le \alpha(t)\sqrt{x^2+y^2} + \beta(t) \tag{1.7.8}$$

for suitable $\alpha, \beta \in C(I)$; the term $\sqrt{x^2+y^2}$ in (1.7.8) can of course be replaced by $|x| + |y|$.

2. The functions appearing in equations (ii) and (iii) of Example 1.5.4 are respectively

$$f(t,x,y) = t\sin(xy) \quad \text{and} \quad f(t,x,y) = \frac{(\sin x)y}{\sqrt{1-t^2}}$$

and evidently satisfy the sublinearity condition (1.7.8); the maximal solutions are thus defined on $\mathbb{R}$ and $]-1,1[$, respectively.

Section 1.6

Exercise 6.1

Two independent solutions of the homogeneous equation $x'' - x = 0$ are

$$v_1(t) = e^t, \quad v_2(t) = e^{-t},$$

and their Wronskian determinant is $W(t) = -2$.

To find a solution of the given equation $x'' - x = 1/(1+e^t)$ we use formula (1.6.13), which yields in our case

$$z(t) = e^t \int \frac{-e^{-t}}{(-2)} \frac{1}{1+e^t} dt + e^{-t} \int \frac{e^t}{(-2)} \frac{1}{1+e^t} dt.$$

Thus, we need solve the two integrals

$$c(t) \equiv \int \frac{1}{e^t(1+e^t)} dt$$

and

$$d(t) \equiv \int \frac{e^t}{1+e^t} dt.$$

The latter can be solved immediately, as we have

$$\int \frac{e^t}{1+e^t} dt = \int \frac{d(1+e^t)}{1+e^t} = \ln(1+e^t) + K.$$

As to the former, we first put $y = e^t$ – so that $dy = ydt$ – to obtain

$$\int \frac{1}{e^t(1+e^t)} dt = \int \frac{1}{y(1+y)} \frac{1}{y} dy.$$

Now we use the decomposition of fractions into simpler fractions to write

$$\frac{1}{y^2(1+y)} = \frac{A}{y} + \frac{B}{y^2} + \frac{C}{1+y} = -\frac{1}{y} + \frac{1}{y^2} + \frac{1}{1+y}.$$

This yields

$$\int \frac{1}{y^2(1+y)} dy = -\ln y - \frac{1}{y} + \ln(1+y) + H,$$

so that

$$c(t) = \int \frac{1}{e^t(1+e^t)} dt = -t - e^{-t} + \ln(1+e^t) + H.$$

Choosing $H = K = 0$, we finally have

$$z(t) = \frac{e^t}{2} c(t) - \frac{e^{-t}}{2} d(t) = \frac{e^t}{2}[-t - e^{-t} + \ln(1+e^t)] - \frac{e^{-t}}{2} \ln(1+e^t).$$

E2. Further exercises

Exercise 1.7.2. Obtain the explicit solutions in Example 1.1.2 following the path indicated in the text itself of the example.

(i) $x' = 3x^{2/3}$

$x = x(t)$ being a solution of (i) means that $x'(t) = 3(x(t))^{2/3}$ for all t in some interval J. Therefore,

$$\frac{x'(t)}{(x(t))^{2/3}} = 3 \quad (t \in J)$$

provided that $x(t) \neq 0$ for $t \in J$. Integrating both members of the above equality we obtain

$$\int \frac{x'(t)}{(x(t))^{2/3}} dt = \int 3\, dt = 3t + C,$$

whence, making in the first integral the substitution $x(t) = y$ and recalling that $\int y^\alpha dy = \frac{y^{\alpha+1}}{\alpha+1}$ for each real $\alpha \neq -1$, we get (putting for convenience $C = 3k$)

$$3(x(t))^{1/3} = 3t + 3k,$$

which yields the solutions

$$x(t) = (t+k)^3 \quad (k \in \mathbb{R}).$$

It is clear that the solution $x \equiv 0$ must be considered separately.

(ii) $x' = -2tx^2$

Proceeding as before, we write

$$\int \frac{x'(t)}{(x(t))^2}\,dt = \int (-2t)\,dt = -t^2 + C.$$

Since $\int 1/y^2 = -1/y$, we then have (putting $C = -k$)

$$-\frac{1}{x(t)} = -t^2 - k$$

and we conclude with the formula

$$x(t) = \frac{1}{t^2 + k},$$

which yields all solutions of (ii) except the trivial one $x \equiv 0$.
(iii) $x' = x(1 - x)$

Here we have two "trivial" solutions, $x \equiv 0$ and $x \equiv 1$.

Consider the more general form $x' = ax(1 - bx)$ of (iii) (with $a, b > 0$). We check to have the solutions

$$x(t) = \frac{Ke^{at}}{1 + Kbe^{at}} \quad (K \in \mathbb{R}). \tag{1.7.9}$$

Indeed, using the identity

$$\frac{1}{ax(1 - bx)} = \frac{1}{ax} + \frac{b}{a}\frac{1}{(1 - bx)}$$

we see that a solution $x = x(t)$ must satisfy the equality

$$\int \frac{x'(t)}{x(t)}\,dt + b\int \frac{x'(t)}{1 - bx(t)}\,dt = \int a\,dt = at + C,$$

which leads to

$$\ln|x(t)| - \ln|1 - bx(t)| = \ln\frac{|x(t)|}{|1 - bx(t)|} = at + C,$$

whence

$$\frac{x(t)}{1 - bx(t)} = Ke^{at} \quad (K \in \mathbb{R}). \tag{1.7.10}$$

We can now recover $x(t)$ from (1.7.10), obtaining (1.7.9).

Exercise 1.7.3. Solve the IVPs

$$\begin{cases} x' = -4t^3x^2 \\ x(0) = -1 \end{cases} \quad \text{and} \quad \begin{cases} x' = \frac{x}{t} + 3t^2e^t \\ x(1) = 0 \end{cases}$$

and for each of them check if the (maximal) solution is global or not.

- The answer to the second problem is *a priori* immediate, as it concerns a linear equation. Using the solution formula (1.1.2), we find explicitly

$$x(t) = 3(t^2 - t)e^t.$$

- As to the first problem, since $f(t,x) = -4t^3x^2$ is defined on $\mathbb{R}^2 = \mathbb{R} \times \mathbb{R}$, it follows that global solutions of the given equation are those defined on the whole of $\mathbb{R}$. Now, except for the trivial solution $x \equiv 0$, the other solutions are obtained writing

$$\int \frac{x'(t)}{x^2(t)}\,dt = -4\int t^3\,dt = -t^4 + K,$$

whence

$$-\frac{1}{x(t)} = -t^4 - H \quad (H = -K),$$

which finally yields

$$x(t) = \frac{1}{t^4 + H}.$$

This shows that only the solutions with $H > 0$ are global. For $H = 0$ we have two maximal solutions (differing only for the interval of definition, respectively $J_1 =]-\infty, 0[$ and $J_2 =]0, +\infty[$) and similarly for $H < 0$ we have three maximal solutions. The (unique) maximal solution of the given Cauchy problem is

$$x(t) = \frac{1}{t^4 - 1}, \quad t \in J =]-1, 1[,$$

and is therefore not global.

Systems of linear equations with constant coefficients (Section 1.4)

Here we give a few simple examples (some of them taken from Apostol's book *Calculus* [4]) that illustrate formula (1.4.19) giving the general solution in terms of the eigenvalues and eigenvectors of the matrix coefficients in the case that A is diagonable in the real field. (Note: as already done in the text, when not necessary we will not distinguish between a vector of $\mathbb{R}^n$ being displayed as a row vector or as a column vector.)

Exercise 1.7.4. Consider the system

$$\begin{cases} x' = 5x + 4y \\ y' = x + 2y. \end{cases} \tag{1.7.11}$$

The eigenvalues of the coefficient matrix A are given by the equation

$$\det(A - \lambda I) = \det \begin{pmatrix} 5-\lambda & 4 \\ 1 & 2-\lambda \end{pmatrix} = (5-\lambda)(2-\lambda) = \lambda^2 - 7\lambda - 4 = 0,$$

yielding $\lambda_1 = 6, \lambda_2 = 1$. The eigenvectors corresponding to λ_1 are the non-zero solutions of the equation $(A - 6I)X = 0$ (I being the identity matrix), that is,

$$\begin{pmatrix} -1 & 4 \\ 1 & -4 \end{pmatrix} \begin{pmatrix} x \\ y \end{pmatrix} = \begin{pmatrix} 0 \\ 0 \end{pmatrix},$$

yielding $y = x/4$. Therefore, we can take $V_1 = (4,1)$ as an eigenvector to λ_1. Similarly we find that $V_2 = (1,-1)$ is as eigenvector associated with λ_2. Applying formula (1.4.19) we thus find

$$X(t) = k_1 e^{\lambda_1 t} V_1 + k_2 e^{\lambda_2 t} V_2 = k_1 e^{6t} \begin{pmatrix} 4 \\ 1 \end{pmatrix} + k_2 e^t \begin{pmatrix} 1 \\ -1 \end{pmatrix} = \begin{pmatrix} 4k_1 e^{6t} + k_2 e^t \\ k_1 e^{6t} - k_2 e^t \end{pmatrix} \equiv \begin{pmatrix} x(t) \\ y(t) \end{pmatrix}$$

as solutions of the system (1.7.11). The solution is made unique by imposing an initial condition; for instance, the solution of (1.7.11) satisfying the condition $X(0) = (x(0), y(0)) = (2,3)$ is given by

$$\begin{cases} x(t) = 4e^{6t} - 2e^t \\ y(t) = e^{6t} + 2e^t. \end{cases}$$

Exercise 1.7.5. Consider now the following slight modification of (1.7.11):

$$\begin{cases} x' = 5x + 4y \\ y' = x + 2y \\ z' = x + 2y + 2z. \end{cases} \qquad \text{Here } A = \begin{pmatrix} 5 & 4 & 0 \\ 1 & 2 & 0 \\ 1 & 2 & 2 \end{pmatrix}, \tag{1.7.12}$$

for which we find the eigenvalues

$$\lambda_1 = 6, \quad \lambda_2 = 1, \quad \lambda_3 = 2$$

and the corresponding eigenvectors

$$V_1 = (8,2,3), \quad V_2 = (1,-1,1), \quad V_3 = (0,0,1).$$

It follows that the general solution of (1.7.12) is

$$\begin{cases} x(t) = 8k_1 e^{6t} + k_2 e^t \\ y(t) = 2k_1 e^{6t} - k_2 e^t \\ z(t) = 3k_1 e^{6t} + k_2 e^t - k_3 e^{2t}, \end{cases} \tag{1.7.13}$$

and for instance the solution that satisfies the initial condition $X(0) = (0,0,1) = \boldsymbol{e}_3$ is $X(t) = (0,0,e^{2t})$.

Exercise 1.7.6. Find the general solutions of the systems

$$(A) \begin{cases} x' = x + y \\ y' = x - y \end{cases} \quad \text{and} \quad (B) \begin{cases} x' = x \\ y' = x + 2y. \end{cases} \tag{1.7.14}$$

Also, find the solution of the system (B) satisfying the initial condition $x(1) = 1, y(1) = 10$.

Exercise 1.7.7. Consider the three matrices

$$A = \begin{pmatrix} 1 & 0 \\ 1 & 1 \end{pmatrix}, \quad B = \begin{pmatrix} 0 & 1 \\ 1 & 0 \end{pmatrix}, \quad C = \begin{pmatrix} 0 & 1 \\ -1 & 0 \end{pmatrix} \tag{1.7.15}$$

and establish which of them is diagonable in the real field.

Exercise 1.7.8. Consider the matrices

$$A = \begin{pmatrix} 1 & 0 & 0 \\ 0 & 1 & 0 \\ 1 & 1 & 2 \end{pmatrix}, \quad B = \begin{pmatrix} 1 & 0 & 0 \\ 1 & 1 & 0 \\ 0 & 0 & 2 \end{pmatrix}$$

and check that A is diagonable, while B is not.

Exercise 1.7.9. The matrix

$$A = \begin{pmatrix} 2 & 1 & 1 \\ 2 & 3 & 2 \\ 3 & 3 & 4 \end{pmatrix}$$

has the eigenvalues

$$\lambda_1 = 1, \quad \lambda_2 = 1, \quad \lambda_3 = 7.$$

(We have repeated λ_1 because of its appearance with algebraic multiplicity 2.)

Show that A is diagonable.

2 Metric and normed spaces

Introduction

The goal of this chapter is that of helping to bridge the gap that a student may find between a basic course of calculus, where one typically learns to manage and compute derivatives and integrals of real-valued functions of one or several real variables and to solve some elementary differential equations, and a course in functional analysis, which one will very likely have to follow in a Master on pure and/or applied mathematics. To do this, he/she must necessarily gain – among others – elementary knowledge of:
- Banach and Hilbert spaces;
- some concrete function spaces.

While Hilbert spaces will be treated in the next chapter in connection with the study of Fourier series, a good common frame for the presentation of the remaining arguments is provided by the study of normed vector spaces; however, this would by far not be sufficient for the study of weak topologies, which is a necessary ingredient of a course in functional analysis and applications to PDEs. Therefore, we choose the context of metric spaces (Section 2.2) as a good intermediate step between general topological spaces and normed vector spaces; this should provide a quite general and flexible frame both to serve the scope of this book and to prepare the reader to make further steps on his/her own account when needed.

A beautiful illustration of the abstract, but intuitive, language of metric spaces – which is largely taken from that of the familiar Euclidean space – and in particular of the idea of metric completeness is given for instance by the contraction mapping theorem (Theorem 2.2.4), one of the simplest, yet most useful, fixed point theorems in analysis.

In order to appreciate further the quite general context of metric spaces, we include in this chapter – besides the strictly necessary Sections 2.3 and 2.4 – two final sections, devoted to compactness (Section 2.5) and connectedness (Section 2.6), with the aim of giving not only the basic facts about these two important topological concepts, but also more general versions of two of the most famous and useful theorems in analysis, both known in their simplest form to first year mathematics students: the Weierstrass theorem on the achievement of

$$\min_{x\in[a,b]} f(x), \qquad \max_{x\in[a,b]} f(x)$$

for a continuous real-valued function f on a bounded closed interval $[a,b]$ and the "intermediate value theorem," which asserts that any continuous function that changes sign on an interval must necessarily have a zero.

On the other hand, as a motivation for the study of metric spaces we find it good to start this chapter dealing concretely with the convergence of sequences and series of

https://doi.org/10.1515/9783111302522-002

functions (Section 2.1). Our order of presentation aims at rendering more transparent the concept of uniform convergence for them in the light of the distance

$$d(f,g) = \sup_{x \in A} |f(x) - g(x)| \quad (*)$$

that equips the space $(B(A), d)$ of bounded real functions defined on A; see Sections 2.3 and 2.4. This also makes clear, via the alternative idea of convergence in the mean, induced by the distance

$$d_1(f,g) = \int_a^b |f(x) - g(x)| \, dx,$$

that we can have on the same function space (in this example, $C([a,b])$), and generally speaking on the same set, two or many different distances, and we are therefore stimulated to study the connections between the various types of convergence.

2.1 Sequences and series of functions. Uniform convergence

A sequence (f_n) of real-valued functions defined in a set A is said to **converge pointwisely** in A if for every $x \in A$, the numerical sequence $(f_n(x))$ is convergent. In this case, the function $f : A \to \mathbb{R}$ defined putting

$$f(x) = \lim_{n\to\infty} f_n(x) \quad (x \in A)$$

is called the **pointwise limit** function of the sequence (f_n).

Exercise 2.1.1. For $n \in \mathbb{N}$ and $x \in \mathbb{R}$ put

$$f_n(x) = \arctan nx, \quad f_n(x) = \frac{\sin^2 nx}{1 + nx^2}, \quad f_n(x) = \frac{nx}{1 + e^{nx}}.$$

For each of the three cases, compute the *set of convergence* of the sequence (f_n) (that is, the set $A = \{x \in \mathbb{R} : (f_n(x)) \text{ converges}\}$) and the pointwise limit function of the sequence.

Exercise 2.1.1 shows, among others, that the pointwise limit of a sequence of continuous functions need *not* be continuous itself. To understand this fact, the concept of *uniform convergence* of a sequence of functions is crucial, and we are now going to introduce it.

Bounded functions

Let A be any set. A real-valued function f defined on A is said to be *bounded* if its image $f(A) = \{f(x) : x \in A\}$ is a bounded subset of $\mathbb{R}$, that is, if there exists a constant k such that $|f(x)| \leq k$ for all $x \in A$. Therefore, for such an f, the number

$$\sup_{x\in A}|f(x)| = \min\{k \in \mathbb{R} : |f(x)| \le k \ \forall x \in A\}$$

is well defined, and it is characterized by the following two properties:

$$|f(x)| \le \sup_{x\in A}|f(x)| \quad \forall x \in A, \tag{2.1.1}$$

$$(|f(x)| \le k \ \forall x \in A) \Rightarrow \sup_{x\in A}|f(x)| \le k. \tag{2.1.2}$$

Definition 2.1.1. Let A be any set. A sequence (f_n) of real-valued, bounded functions defined in A is said to **converge uniformly** to the (bounded) function $f : A \to \mathbb{R}$ if

$$\lim_{n\to\infty} \sup_{x\in A}|f_n(x) - f(x)| = 0.$$

Clearly, if (f_n) converges uniformly to f, then it converges pointwisely to f, for

$$|f_n(x) - f(x)| \le \sup_{x\in A}|f_n(x) - f(x)| \quad (x \in A, n \in \mathbb{N}).$$

This – together with the uniqueness of the limit of real sequences – implies in particular that the uniform limit function of (f_n) (when it exists) is unique and coincides with its pointwise limit.

Example 2.1.1. The sequence $(\arctan nx)$ converges pointwisely, but not uniformly, on $\mathbb{R}$ to the function

$$f(x) = \begin{cases} \frac{\pi}{2} & x > 0 \\ 0 & x = 0 \\ -\frac{\pi}{2} & x < 0. \end{cases}$$

Indeed,

$$\sup_{x\in\mathbb{R}}|f_n(x) - f(x)| = \frac{\pi}{2} \quad \forall n \in \mathbb{N}.$$

Example 2.1.2. The sequence $\frac{nx}{n^2+x^2}$ converges uniformly to 0 on $[0,1]$, for

$$\sup_{[0,1]} \frac{nx}{n^2 + x^2} \le \frac{1}{n}.$$

On the other hand, the sequence $\frac{nx}{1+n^2x^2}$ converges pointwisely, but not uniformly, to 0 on $[0,1]$, because

$$\sup_{[0,1]} \frac{nx}{1 + n^2x^2} = \max_{[0,1]} \frac{nx}{1 + n^2x^2} = \frac{1}{2}, \quad \text{for all } n \in \mathbb{N}.$$

The following statement shows the importance of uniform convergence as far as the continuity of the limit function is concerned.

Theorem 2.1.1. *Let (f_n) be a sequence of real-valued, bounded functions defined in $A \subset \mathbb{R}$ and converging uniformly to the (bounded) function $f : A \to \mathbb{R}$. If each f_n is continuous in A, then so is the limit function f.*

The proof of Theorem 2.1.1 could be given right now, but we prefer to postpone it for a while (see Theorem 2.3.2) because it gains clarity and conciseness in the context of metric function spaces. On the other hand, when applied in case A is an **interval**, Theorem 2.1.1 implies in turn two more basic statements concerning the limit function of a sequence: the first is known as **passage to the limit under the sign of integral** and the second as **passage to the limit under the sign of derivative**.

Theorem 2.1.2. *Let (f_n) be a sequence of real-valued, continuous functions defined in the interval $[a, b]$ and converging uniformly to the function $f : [a, b] \to \mathbb{R}$. Then*

$$\lim_{n\to\infty} \int_a^b f_n(x)\,dx = \int_a^b f(x)\,dx.$$

Proof. First, f is continuous on the basis of Theorem 2.1.1. Next, for each $n \in \mathbb{N}$ we have

$$\left| \int_a^b f_n(x)\,dx - \int_a^b f(x)\,dx \right| \le \int_a^b |f_n(x) - f(x)|\,dx \le L_n(b-a), \tag{2.1.3}$$

with $L_n = \sup_{x\in[a,b]} |f_n(x) - f(x)|$. Since by assumption $L_n \to 0$ as $n \to \infty$, it follows that the left-hand side of (2.1.3) also does so, proving the result. □

Note. From now on, when speaking of uniform convergence of a sequence (f_n), we tacitly assume that both the f_n's and the limit f are bounded functions.

Theorem 2.1.3. *Let (f_n) be a sequence of real-valued C^1 functions defined on some interval $I \subset \mathbb{R}$. Suppose that:*
(i) *the real sequence $(f_n(x))$ converges for at least one $x \in I$;*
(ii) *the sequence (f_n') of the derivatives converges uniformly on I to a function g.*

Then (f_n) converges pointwisely on I to a limit function f which is itself of class C^1 on I, and we have

$$f'(x) = g(x)$$

*for every $x \in I$. Finally, if I is **bounded**, then the convergence of (f_n) to f is uniform.*

Proof. Let $x_0 \in I$ be such that $(f_n(x_0))$ converges and put $l = \lim_{n\to\infty} f_n(x_0)$. As f_n is of class C^1, by the fundamental theorem of calculus we can write

$$f_n(x) = f_n(x_0) + \int_{x_0}^{x} f_n'(t)\,dt \quad (x \in I, n \in \mathbb{N}). \tag{2.1.4}$$

Letting $n \to \infty$ in (2.1.4) and using Theorem 2.1.1 for the sequence (f_n') on $[x_0, x]$, we see that $(f_n(x))$ converges for any $x \in I$ (that is, (f_n) converges pointwisely on I) and precisely that

$$f(x) \equiv \lim_{n\to\infty} f_n(x) = l + \int_{x_0}^{x} g(t)\,dt \quad (x \in I). \tag{2.1.5}$$

This shows that $f \in C^1(I)$ and that $f' = g$ in I. Thus, it remains only to show that (f_n) converges uniformly to g in case $I = (a, b)$ is a bounded interval. However, for $x \in I$, if for example $x > x_0$, we have

$$\begin{aligned} |f_n(x) - f(x)| &= \left| f_n(x_0) + \int_{x_0}^{x} f_n'(t)\,dx - l - \int_{x_0}^{x} g(t)\,dt \right| \\ &\le |f_n(x_0) - l| + \int_{x_0}^{x} |f_n'(t) - g(t)|\,dt \\ &\le |f_n(x_0) - l| + H_n(b-a), \end{aligned}$$

with $H_n = \sup_{x\in I} |f_n'(x) - f'(x)|$. Therefore,

$$L_n = \sup_{x\in[a,b]} |f_n(x) - f(x)| \le |f_n(x_0) - l| + H_n(b-a).$$

Letting $n \to \infty$ the result follows. □

The following simplified form of Theorem 2.1.3 is often useful.

Corollary 2.1.1. *Let $(f_n) \subset C^1(I)$ and suppose that as $n \to \infty$*

$$f_n \to f, \quad f_n' \to g \quad \textit{uniformly on } I.$$

Then $f \in C^1(I)$ and $f' = g$.

Series of functions

The importance of uniform convergence can be seen even better by looking at *series* of functions. Let again (f_n) be a sequence of real-valued functions defined in A and suppose that the numerical series $\sum_{n=1}^{\infty} f_n(x)$ converges for each $x \in A$ – in other words, that $\sum_{n=1}^{\infty} f_n$ converges pointwisely on A. Then we can consider the sum function of the series, defined by putting

$$f(x) = \sum_{n=1}^{\infty} f_n(x) \quad (x \in A), \tag{2.1.6}$$

and ask as above if, for instance, f is continuous (or C^k) if so are all f_n. This problem is of high interest in practice, partly because (except for a few special cases) we do *not* know explicitly the function f and partly because in many applications we obtain the solution of a problem – for example, of a differential equation – precisely in the form (2.1.6), where f_n ($n \in \mathbb{N}$) are "elementary" solutions of the problem.

Example 2.1.3 (The Dirichlet problem for the Laplace equation in a rectangle). Let $R = \{(x,y) \in \mathbb{R}^2 : 0 < x < \pi, 0 < y < 1\}$, let $\overline{R} = \{(x,y) \in \mathbb{R}^2 : 0 \le x \le \pi, 0 \le y \le 1\}$, and let $h \in C([0,\pi])$ be such that $h(0) = h(\pi) = 0$. The problem is to find a function $u \in C(\overline{R}) \cap C^2(R)$ – that is, twice continuously differentiable in the open rectangle R and continuous up to the boundary of R – which satisfies the *boundary value problem*

$$\begin{cases} \Delta u = \frac{\partial^2 u}{\partial x^2} + \frac{\partial^2 u}{\partial y^2} = 0 & (x,y) \in R \\ u(\pi,y) = u(x,1) = u(0,y) = 0 & 0 \le x \le \pi, 0 \le y \le 1 \\ u(x,0) = h(x) & 0 \le x \le \pi. \end{cases} \tag{2.1.7}$$

Using separation of variables, a solution u of (2.1.7) can be found in the form

$$u(x,y) = \sum_{n=1}^{\infty} c_n \sin nx \frac{\sinh n(1-y)}{\sinh n}, \tag{2.1.8}$$

where for $z \in \mathbb{R}$ we put $\sinh z = (e^z - e^{-z})/2$ and for $n \in \mathbb{N}$

$$c_n = \frac{2}{\pi} \int_0^{\pi} h(x) \sin nx \, dx.$$

One has to check not only that the definition of u is sensible – i. e., that the series defining u is convergent at any point of $\overline{R}$ – but also that the resulting u has the necessary regularity properties (that is, $u \in C(\overline{R}) \cap C^2(R)$) required to be a solution of (2.1.7); this will be discussed in some detail in Chapter 4, Section 4.4.

Returning in general to the continuity of an f defined as in (2.1.6), this will be granted by Theorem 2.1.1 provided that each f_n is (bounded and) continuous on A and that the series converges uniformly in A, meaning by definition that the sequence (s_n) of the finite sums

$$s_n = \sum_{k=1}^{n} f_k \quad (n \in \mathbb{N})$$

does so.

A useful criterion to check the uniform convergence of a series is the popular **Weierstrass M-test**, which is stated as follows.

Proposition 2.1.1. *Suppose that (f_n) is a sequence of bounded functions defined on a set A and suppose that there exists a sequence (M_n), with $M_n \geq 0$ for all $n \in \mathbb{N}$, such that:*
(i) *$|f_n(x)| \leq M_n$ for all $x \in A$ and all $n \in \mathbb{N}$;*
(ii) *the series $\sum_{n=1}^{\infty} M_n$ is convergent.*

Then the series $\sum_{n=1}^{\infty} f_n$ converges ***uniformly*** *in A.*

The Weierstrass M-test is a simple consequence of a general result (Theorem 2.4.1) that will be given later in this chapter in the context of Banach spaces. However, in order to appreciate its usefulness we shall start employing it in the examples that follow. One thing to observe immediately is that assumptions (i) and (ii) of Proposition 2.1.1 imply convergence of the series

$$\sum_{n=1}^{\infty} f_n(x)$$

for every $x \in A$ – that is, *pointwise* convergence on A of $\sum_{n=1}^{\infty} f_n$ – by virtue of the comparison test for numerical series.

We now state for further use the specific form that Theorem 2.1.2 and Theorem 2.1.3, or more precisely its Corollary 2.1.1, take for series of functions.

Corollary 2.1.2 (Term-by-term integration of a series of functions). *Let (f_n) be a sequence of real-valued, continuous functions defined in the interval $[a, b]$. Suppose that the series $\sum_{n=1}^{\infty} f_n$ converges uniformly to the function $f : [a, b] \to \mathbb{R}$. Then*

$$\int_a^b \sum_{n=1}^{\infty} f_n(x)\, dx = \sum_{n=1}^{\infty} \int_a^b f_n(x)\, dx.$$

Corollary 2.1.3 (Term-by-term derivation of a series of functions). *Let (f_n) be a sequence of real-valued, C^1 functions defined in an interval $I \subset \mathbb{R}$. Suppose that the two series*

$$\sum_{n=1}^{\infty} f_n, \quad \sum_{n=1}^{\infty} f_n'$$

converge uniformly on I to the functions f and g, respectively. Then $f \in C^1(I)$ and $f' = g$; that is,

$$D\left(\sum_{n=1}^{\infty} f_n\right)(x) = \sum_{n=1}^{\infty} Df_n(x).$$

Example 2.1.4. The series

$$\sum_{n=1}^{\infty} \frac{1}{\sqrt{n+\ln n}} \ln\left(\frac{1+nx}{nx}\right)$$

converges for any $x > 0$, and its sum function is C^1 on $]0, \infty[$. Indeed, if a is any positive number, we have for $x \geq a$

$$\frac{1}{\sqrt{n+\ln n}} \ln\left(1+\frac{1}{nx}\right) \leq \frac{1}{\sqrt{n}} \ln\left(1+\frac{1}{na}\right) \sim \frac{1}{n^{3/2}},$$

where $a_n \sim b_n$ means that $\lim_{n\to\infty} a_n/b_n = 1$. As to the derivatives, note that for $x \geq a$

$$\left|D\left(\ln\left(1+\frac{1}{nx}\right)\right)\right| = \frac{1}{1+\frac{1}{nx}} \frac{1}{nx^2} \leq \frac{1}{na^2},$$

showing that also the series $\sum_{n=1}^{\infty} |f_n'(x)|$ is uniformly dominated by $\sum_{n=1}^{\infty} \frac{1}{n^{3/2}}$. Thus, by the Weierstrass M-test, both series $\sum_{n=1}^{\infty} f_n$ and $\sum_{n=1}^{\infty} f_n'$ converge uniformly in $[a, \infty[$. As this holds for any $a > 0$, it follows that the sum function is C^1 on $]0, \infty[$.

Example: power series

A **power series** is a series of the form

$$\sum_{0}^{\infty} a_n (x - x_0)^n, \tag{2.1.9}$$

where $x_0 \in \mathbb{R}$ is called the *center* of the series and (a_n) is a sequence of real numbers. (To be better understood, power series should indeed be considered in the complex plane, where they are at the basis of the theory of analytic functions of a complex variable; see for instance Rudin [6].) Putting $y = x - x_0$, we see that it is enough to consider power series of the form

$$\sum_{0}^{\infty} a_n x^n \tag{2.1.10}$$

having center $x_0 = 0$. The basic example is, of course, the **geometric series** $\sum_0^{\infty} x^n$, which we know, by explicit calculation of its nth partial sum,

$$s_n = \sum_{0}^{n-1} x^i = 1 + x + \cdots + x^{n-1} = \frac{1-x^n}{1-x},$$

to converge exactly for those x belonging to the interval $]-1, 1[$. The special feature of power series is precisely that kind of behavior; indeed, we are going to show that – except if the series converges only for $x = 0$ (for example, $\sum_0^{\infty} n^n x^n$) or for every $x \in \mathbb{R}$

(for example, $\sum_0^\infty \frac{x^n}{n!}$) – there exists a number $R > 0$ such that (2.1.10) converges if $|x| < R$ and does not converge if $|x| > R$; the behavior at the points $x = \pm R$ must be checked in each case. To prove this statement, the key point is the following result.

Theorem 2.1.4. *Suppose that the series $\sum_0^\infty a_n x^n$ converges in a point $x_1 \neq 0$. Then:*
(i) *the series converges absolutely in every point x with $|x| < |x_1|$;*
(ii) *the series converges uniformly on every interval $[-r, r]$ with $r < |x_1|$.*

Proof. (i) Fix an x with $|x| < |x_1|$. We have

$$|a_n x^n| = |a_n||x|^n = |a_n||x_1|^n \left(\frac{|x|}{|x_1|}\right)^n \leq M\left(\frac{|x|}{|x_1|}\right)^n \tag{2.1.11}$$

for some $M > 0$ and all $n \in \mathbb{N}$, because $a_n x_1^n \to 0$ for $n \to \infty$, as a necessary consequence of the assumed convergence of the series $\sum_0^\infty a_n x_1^n$. Thus, the series $\sum_0^\infty |a_n x^n|$ converges, because it is term-by-term majorized by the geometrical series $\sum_0^\infty a_n q^n$, $q \equiv |x|/|x_1| < 1$.

(ii) Work as before improving inequality (2.1.11) to

$$|a_n x^n| \leq M\left(\frac{|x|}{|x_1|}\right)^n \leq M\left(\frac{r}{|x_1|}\right)^n \tag{2.1.12}$$

holding for all $x \in [-r, r]$ and all $n \in \mathbb{N}$, and then use the Weierstrass M-test (Proposition 2.1.1) with $M_n \equiv M(r/|x_1|)^n$. □

Remark 2.1.1. It follows from Theorem 2.1.4 that if the series (2.1.10) does not converge in a point x_2, then it cannot converge in any point x with $|x| > |x_2|$.

Proposition 2.1.2. *Suppose that the series* (2.1.10) *converges at some $x = x_1 \neq 0$ and does not converge at some $x = x_2$. Then there exists a real number $R > 0$ such that:*
- (2.1.10) *converges if $|x| < R$;*
- (2.1.10) *does not converge if $|x| > R$.*

Proof. Let

$$S \equiv \left\{x \in \mathbb{R} : \sum_0^\infty a_n x^n \text{ converges}\right\} \tag{2.1.13}$$

and put $R \equiv \sup_{x \in S} |x|$ (note that by our assumptions, S is non-empty and bounded, for necessarily $|x| \leq |x_2|$ if $x \in S$; see Remark 2.1.1). Now simply use the properties of the supremum to prove both statements in Proposition 2.1.2 (in particular, if $|x| < R$, there must be an $x_1 \in S$ such that $|x| < |x_1|$; then use Theorem 2.1.4). □

Definition 2.1.2. The *radius of convergence* of the power series (2.1.10) is the extended non-negative number R defined as follows:

- $R = 0$ if (2.1.10) converges only for $x = 0$;
- $R = +\infty$ if (2.1.10) converges for every $x \in \mathbb{R}$;
- $R \equiv \sup_{x \in S}$ (with S as in (2.1.13)) in the other cases.

The radius of convergence can be explicitly **computed** if we know the coefficients a_n and especially their asymptotic behavior: this, together with further information on power series, is shortly indicated in the final section of this chapter. An excellent reference for a complete discussion of power series is, for instance, Rudin's book [6]; for an even more general approach, see Dieudonné [7].

2.2 Metric spaces and continuous functions

Distance. Open sets. Interior of a set

Definition 2.2.1. Let X be a non-empty set. A *distance* (or *metric*) on X is a function $d : X \times X \to \mathbb{R}$ satisfying the following properties:
(i) $d(x,y) \geq 0$ for all $x, y \in X$ and $d(x,y) = 0$ if and only if $x = y$;
(ii) $d(x,y) = d(y,x)$ for all $x, y \in X$;
(iii) $d(x,y) \leq d(x,z) + d(z,y)$ for all $x, y, z \in X$ ("triangle inequality").

A *metric space* is a pair (X, d), where d is a distance on X.

Example 2.2.1. $X = \mathbb{R}^n$ is a metric space endowed with the usual *Euclidean distance*

$$d(x,y) = \|x - y\| = \sqrt{\sum_{i=1}^{n} (x_i - y_i)^2} \tag{2.2.1}$$

between the points $x = (x_1, \ldots x_n)$ and $y = (y_1, \ldots, y_n)$ of $\mathbb{R}^n$. If $n = 1$, (2.2.1) reduces to the *absolute value* $|x - y|$ of $x - y$.

Example 2.2.2. More generally, let E be a real vector space. Suppose E is equipped with a *norm*, that is, a function $\|.\|$ from E to $\mathbb{R}$, satisfying the following properties:
(i) $\|x\| \geq 0$ for all $x \in \mathbb{E}$, and $\|x\| = 0$ iff $x = 0$;
(ii) $\|\lambda x\| = |\lambda|\|x\|$ for all $\lambda \in \mathbb{R}$ and all $x \in E$;
(iii) $\|x + y\| \leq \|x\| + \|y\|$ for all $x, y \in E$.

Then E is called a *normed vector space*, and it is easily checked that the function $d : E \times E \to \mathbb{R}$ defined by putting

$$d(x,y) \equiv \|x - y\| \quad (x, y \in E) \tag{2.2.2}$$

is a distance on E, which is said to be *induced* by the norm.

Exercise 2.2.1. Consider the function space $C([a,b])$ and show that the usual algebraic operations of pointwise sum and product by a real number make it into a vector space, which can be normed putting

$$\|x\|_1 \equiv \int_a^b |x(t)|\, dt. \tag{2.2.3}$$

The distance induced by this norm is therefore

$$d_1(x,y) = \int_a^b |x(t) - y(t)|\, dt. \tag{2.2.4}$$

Exercise 2.2.2. Let X be any set and let $d : X \times X \to \mathbb{R}$ be defined as follows:

$$\begin{cases} d(x,y) = 1 & \text{if } x \neq y \\ d(x,y) = 0 & \text{if } x = y. \end{cases} \tag{2.2.5}$$

Show that d is a distance on X.

In the following definitions, we assume that (X,d) is a metric space.

Definition 2.2.2. For $x \in X$ and $r > 0$, the set

$$B(x,r) = \{y \in X : d(y,x) < r\}$$

is called the *ball* (or *spherical neighborhood*) of center x and radius r.

Note that $B(x,r)$ is non-empty as it contains x; however, it may be reduced to $\{x\}$ itself (see Exercise 2.2.2). If $X = \mathbb{R}$, then $B(x,r)$ is the interval $]x-r, x+r[$; if $X = \mathbb{R}^2$ and $z = (x_0, y_0)$, then $B(z,r)$ is the circle defined by the inequality

$$(x - x_0)^2 + (y - y_0)^2 < r^2.$$

Definition 2.2.3. Let A be a non-empty subset of X. A point $x \in A$ is said to be *interior to* A if $B(x,r) \subset A$ for some $r > 0$. A is said to be *open* if each of its points is interior to A.

The whole space X is evidently open; the empty set is open by definition.

Exercise 2.2.3. The union of any family of open sets is open. The intersection of a *finite* family of open sets is open.

Exercise 2.2.4. A ball is an open set. Show that for each $x \in B(x_0, r)$, one has $B(x, r(x)) \subset B(x_0, r)$, where $r(x) = r - d(x, x_0) > 0$.

Exercise 2.2.5. The interval $]a, +\infty[= \{x \in \mathbb{R} : x > a\}$ is open in $\mathbb{R}$.

Given a subset A of X, the set of all points which are interior to A is called the *interior* of A and is denoted int A. We have $intA \subset A$, and it follows by Definition 2.2.3 that $intA = A$ if and only if A is open.

Exercise 2.2.6. Show that for any $A \subset X$, $intA$ is an open set (use Exercise 2.2.4).

Remark 2.2.1. If X is a metric space with distance d, any non-empty subset F of X becomes itself a metric space with distance $d_F \equiv d_{|F\times F}$ (restriction of d to $F \times F$); this distance is said to be *induced* on F by that of X. Thus, it makes perfectly sense to speak, for instance, of the metric space $[0, 1]$ (where, without further mention, the distance is that induced by the Euclidean distance on $\mathbb{R}$).

Convergent sequences and Cauchy sequences. Complete metric spaces

Definition 2.2.4. A sequence (x_n) of points of X is said to *converge to the point* $x_0 \in X$ if given any $\epsilon > 0$, there exists an integer n_0 such that $d(x_n, x_0) < \epsilon$ for all $n \geq n_0$.

In other words, (x_n) converges to x_0 if any given neighborhood of x_0 contains all x_n except for a finite number of indices. It follows by the properties of the distance d on X that there is at most one point x_0 to which a sequence (x_n) can converge; it is called the *limit* of the sequence (x_n), and in this case we write

$$\lim_{n\to\infty} x_n = x_0 \quad \text{or} \quad x_n \to x_0.$$

Definition 2.2.5. A sequence (x_n) of points of X is said to be *a Cauchy sequence* if given any $\epsilon > 0$, there exists an integer n_0 such that $d(x_n, x_m) < \epsilon$ for all $n, m \geq n_0$.

Exercise 2.2.7. Any convergent sequence is a Cauchy sequence.

Definition 2.2.6. A metric space (X, d) is said to be *complete* if any Cauchy sequence in X converges to a point of X.

Example 2.2.3. The Euclidean space $\mathbb{R}^n$ (see Example 2.2.1) is complete. The set $\mathbb{Q}$ of rational numbers, equipped with the distance inherited by $\mathbb{R}$, is not complete.

Example 2.2.4. Let $X = C([a, b])$ with the distance induced by the norm

$$\|x\| = \max_{t\in[a,b]} |x(t)|.$$

As will be shown – in a more general context – in Corollary 2.3.1, this is a complete metric space. On the other hand, if $C([a, b])$ is equipped with the integral norm considered in Exercise 2.2.1, the resulting metric space is *not* complete.

Continuous functions between metric spaces

Definition 2.2.7. Let (X, d) and (Y, d') be metric spaces and let $f : X \to Y$. We say that f is **continuous at the point** $x_0 \in X$ if given any $\epsilon > 0$, there exists an $r > 0$ such that for every $x \in X$ with $d(x, x_0) < r$ we have

$$d'(f(x), f(x_0)) < \epsilon.$$

In terms of neighborhoods, this is the same as saying that given $\epsilon > 0$, there is an $r > 0$ such that $f(B(x_0, r)) \subset B(f(x_0), \epsilon)$.

By virtue of Remark 2.2.1, the same definition applies with the obvious changes if f is only defined on a subset A of X.

Theorem 2.2.1. *Let X, Y, Z be three metric spaces and let $f : X \to Y$ and $g : Y \to Z$. If f is continuous at $x_0 \in X$ and g is continuous at $f(x_0)$, then $g \circ f$ is continuous at x_0.*

Proof. The proof is the same as that of Theorem 0.0.1 given in the Preliminaries. □

Theorem 2.2.2. *Let X and Y be metric spaces and let $f : X \to Y$. Then f is continuous at $x_0 \in X$ if and only if for* any *sequence $(x_n) \subset X$ convergent to x_0, the sequence $(f(x_n))$ converges to $f(x_0)$.*

Proof. (a) Suppose that f is continuous at $x_0 \in X$, and given $\epsilon > 0$ let $r > 0$ be such that for every $x \in B(x_0, r)$ we have $f(x) \in B(f(x_0), \epsilon)$. Take a sequence $(x_n) \subset X$ converging to x_0 and let $n_0 \in \mathbb{N}$ be such that $x_n \in B(x_0, r)$ for $n \geq n_0$; then it follows that $f(x_n) \in B(f(x_0), \epsilon)$ for $n \geq n_0$, proving the result.

(b) Assume that $f(x_n) \to f(x_0)$ whenever $x_n \to x_0$ and suppose by way of contradiction that f is **not** continuous at x_0. Thus, there is an $\epsilon_0 > 0$ such that, for any $\delta > 0$, we can find an $x \in B(x_0, \delta)$ such that $f(x) \notin B(f(x_0), \epsilon_0)$. Taking $\delta_n = \frac{1}{n}$ $(n \in \mathbb{N})$, we construct a sequence (x_n) that converges to x_0, but such that $f(x_n) \notin B(f(x_0), \epsilon_0)$ for any $n \in \mathbb{N}$. In particular, $(f(x_n))$ cannot converge to $f(x_0)$, contradicting our assumption and thus proving the result. □

Example 2.2.5. Let $X = C([a, b])$ normed as in Example 2.2.4 and let $F : X \to \mathbb{R}$ be defined as follows:

$$F(x) = \int_a^b x(t)\,dt \quad (x \in X).$$

If we take any two functions $x, y \in X$, we have

$$|F(x) - F(y)| = \left|\int_a^b (x(t) - y(t))\,dt\right| \leq \int_a^b |x(t) - y(t)|\,dt \leq (b - a)\|x - y\|,$$

which shows that F is not only continuous, but indeed *Lipschitz* continuous (see Definition 2.2.9 below). By virtue of Theorem 2.2.2, it follows that if $(x_n) \subset X$ converges in X – that is, uniformly on $[a, b]$, see Definition 2.1.1 – to $x \in X$, then $F(x_n) \to F(x)$. We thus find the content of Theorem 2.1.2.

A function $f : X \to Y$ is said to be **continuous** if it is continuous at every point of X. As shown by the next theorem, such a *global* continuity can be characterized via the **preimages**

$$f^{-1}(V) \equiv \{x \in X : f(x) \in V\}$$

through f of the open sets $V \subset Y$.

Theorem 2.2.3. *Let X and Y be metric spaces and let $f : X \to Y$. Then f is continuous if and only if for* any *open set $V \subset Y$, $f^{-1}(V)$ is an open set in X.*

Proof. (a) Suppose that f is continuous and let $V \subset Y$ be open. Pick any point $x_0 \in f^{-1}(V)$; we need to show that x_0 is interior to $f^{-1}(V)$. However, as $f(x_0) \in V$ and V is open, there is an $\epsilon > 0$ so that $B(f(x_0), \epsilon) \subset V$. Moreover, the continuity of f at x_0 shows that there is an $r > 0$ so that $f(B(x_0, r)) \subset B(f(x_0), \epsilon)$. Therefore, $f(B(x_0, r)) \subset V$, which is equivalent to saying that $B(x_0, r) \subset f^{-1}(V)$.

(b) Suppose that for any open subset V of Y, $f^{-1}(V)$ is open in X. Let $x_0 \in X$ and $\epsilon > 0$ be given; then as $B(f(x_0), \epsilon)$ is open (Exercise 2.2.4) our assumption implies that

$$U \equiv f^{-1}(B(f(x_0), \epsilon))$$

is open in X, and evidently $x_0 \in U$. Thus, there is a ball $B(x_0, r) \subset U$, and by the definition of U this is equivalent to saying that $f(B(x_0), r) \subset B(f(x_0), \epsilon)$, q.e.d. □

Uniformly continuous and Lipschitz continuous functions

Definition 2.2.8. Let X and Y be metric spaces with distances d and d', respectively, and let $f : X \to Y$. We say that f is **uniformly continuous** on X if given any $\epsilon > 0$, there exists an $r > 0$ such that for every $x, y \in X$ with $d(x, y) < r$ we have

$$d'(f(x), f(y)) < \epsilon.$$

It is clear that a uniformly continuous function is *a fortiori* continuous (roughly speaking, we say that in the former case "given ϵ, the corresponding r does not depend on the point x_0"). The reverse statement is not true in general; see for instance Example 2.5.2 in this chapter.

Exercise 2.2.8. Prove that if $f : X \to Y$ is uniformly continuous, then for any Cauchy sequence $(x_n) \subset X$, $(f(x_n))$ is a Cauchy sequence in Y.

A further strengthening of the continuity property is obtained considering the *Lipschitz continuous* functions, which we define next.

Definition 2.2.9. Let X and Y be metric spaces with distances d and d', respectively, and let $f : X \to Y$. We say that f is **Lipschitz continuous** (or **Lipschitzian**) on X if there exists a constant $k \geq 0$ such that

$$d'(f(x), f(y)) \leq k d(x, y) \tag{2.2.6}$$

for every $x, y \in X$.

Exercise 2.2.9. Check that if $f : X \to Y$ is Lipschitzian, then it is uniformly continuous.

When it is useful, an f satisfying (2.2.6) is said to be Lipschitzian **of constant** k. The importance of the best constant k for which a condition like (2.2.6) holds is seen by one of the most famous theorems in analysis, the contraction mapping theorem, which we state and prove below. Before that, let us consider in the next examples two important classes of functions satisfying the Lipschitz condition.

Example 2.2.6. Let $I = [a, b]$ be a closed bounded interval of $\mathbb{R}$ and suppose that $f : I \to \mathbb{R}$ is of class C^1. Then f is Lipschitzian in I. Indeed, if we take any two points $x, y \in I$ we have

$$f(x) - f(y) = f'(c)(x - y)$$

for some c between x and y (Lagrange's theorem). As f' is bounded by the Weierstrass theorem (Theorem 0.0.3), it follows that for some $k > 0$, we have

$$|f(x) - f(y)| \leq k|x - y| \quad \forall x, y \in I.$$

Note that the assumptions made on f are in fact stronger than necessary: irrespective of the interval I (closed or not, bounded or not), what we really need is that the derivative of f be **bounded** in I.

The importance of this example lies in the fact that using differential calculus for functions of several variables, it can be easily extended to such functions, with some restriction on the sets to be considered. For instance, if $A \subset \mathbb{R}^n$ is open and **convex** (meaning that for any two points $\mathbf{x}, \mathbf{y} \in A$ the whole segment joining them stays in A) and $f : A \to \mathbb{R}$ is of class C^1, then the mean value theorem (Theorem 0.0.7 in the Preliminaries) tells us that given any two points $\mathbf{x}, \mathbf{y} \in A$ we have

$$f(\mathbf{x}) - f(\mathbf{y}) = \nabla f(\mathbf{z}) \cdot (\mathbf{x} - \mathbf{y})$$

for some $\mathbf{z}$ on the segment joining them. Then – using the Cauchy–Schwarz inequality – it follows as before that if $\|\nabla f(\mathbf{z})\| \leq K$ for some $K \geq 0$ and for all $\mathbf{z} \in A$ (that is to say, if the partial derivatives of f are all bounded in A), then

$$|f(\mathbf{x}) - f(\mathbf{y})| \leq K\|\mathbf{x} - \mathbf{y}\| \quad \forall \mathbf{x}, \mathbf{y} \in A.$$

Example 2.2.7 (Bounded linear operators). Let E, F be real vector spaces and let $T : E \to F$ be a linear map. T is said to be **bounded** if there exists a constant $k \geq 0$ such that

$$\|T(x)\| \leq k\|x\| \quad \forall x \in E. \tag{2.2.7}$$

This property means that T maps bounded subsets of E into bounded subsets of F; this justifies the name (though it contrasts with the usual meaning of bounded function). The linearity of T immediately proves that then T is Lipschitzian of constant k:

$$\|T(x) - T(y)\| = \|T(x - y)\| \le k\|x - y\| \quad \forall x, y \in E.$$

On the other hand, it is fundamental to note that if a linear map $T : E \to F$ is continuous at a point $x_0 \in E$, then it satisfies a condition like (2.2.7) (and is thus Lipschitzian on the whole E, as already said). To see this, just write down the definition of continuity of T at x_0: fixing for instance $\epsilon = 1$, there will be an $r > 0$ such that

$$\|T(x) - T(x_0)\| = \|T(x - x_0)\| \le 1 \quad \text{for all } x \in E \quad \text{with } \|x - x_0\| \le r.$$

(We have used for convenience the symbol $\le$ rather than the strict $<$; it is easy to see that using either of them in Definition 2.2.7 is equivalent.) Making the change of variable $y = x - x_0$, we thus see that $\|T(y)\| \le 1$ if $\|y\| \le r$. However, given any $z \in E, z \neq 0$, putting $y = rz/\|z\|$ we see that $\|y\| = r$ and therefore

$$\left\|T\left(\frac{rz}{\|z\|}\right)\right\| = \frac{r}{\|z\|}\|T(z)\| \le 1 \quad \forall z \in E, z \neq 0,$$

whence we conclude that T satisfies inequality (2.2.7) with $k = 1/r$ (note that for $x = 0$, (2.2.7) is trivially satisfied since $T(0) = 0$). These arguments show that linear maps are quite special in that for them (Lipschitz) continuity, continuity at a single point, and the property (2.2.7) are all equivalent.

Example 2.2.8. The map F considered in Example 2.2.5 is a bounded linear map of $X = C([a, b])$ (with the sup norm) into $\mathbb{R}$. By the same argument we can check that the map G defined on X putting

$$G(x)(t) = \int_a^t x(s)ds \quad (x \in X)$$

is a bounded linear map of X into itself.

Further examples of bounded linear operators are given in the Additions to this chapter.

The contraction mapping theorem

Definition 2.2.10. A mapping $F : X \to X$ is said to be a *contraction* if there exists a constant α, $0 < \alpha < 1$, such that

$$d(F(x), F(y)) \le \alpha d(x, y) \quad \forall x, y \in X. \tag{2.2.8}$$

In other words, a contraction is a Lipschitz function with Lipschitz constant less than 1.

Theorem 2.2.4. *Let X be a complete metric space and let $F : X \to X$ be a contraction. Then there exists a unique $x \in X$ such that $F(x) = x$.*

Remark. For a map $F : X \to X$ (with X any non-empty set), a point $x \in X$ such that $F(x) = x$ is called a **fixed point** of F. Thus, Theorem 2.2.4 says that if X is a complete metric space, then a contraction of X into itself possesses a unique fixed point.

Proof of Theorem 2.2.4

Let x_0 be any point of X and define by recurrence a sequence (x_n) in X on putting $x_1 = F(x_0)$, $x_2 = F(x_1), \ldots$; that is,

$$x_n = F(x_{n-1}), \quad n \in \mathbb{N}. \tag{2.2.9}$$

We claim that (x_n) is a Cauchy sequence. To show this, first use (2.2.8) to prove by induction the inequality

$$d(x_n, x_{n-1}) \le a^{n-1} d(x_1, x_0), \quad n \in \mathbb{N}. \tag{2.2.10}$$

Next, let $s_n = \sum_{k=0}^{n-1} a^k$ be the nth partial sum of the geometric series $\sum_{k=0}^{\infty} a^k$ and observe that

$$d(x_n, x_m) \le d(x_1, x_0)|s_n - s_m|, \quad n, m \in \mathbb{N}. \tag{2.2.11}$$

This follows by the triangle inequality and (2.2.10), for we have (assuming, e. g., that $n > m$)

$$d(x_n, x_m) \le d(x_n, x_{n-1}) + \cdots d(x_{m+1}, x_m). \tag{2.2.12}$$

Inequality (2.2.11), together with the convergence of the series $\sum_{k=0}^{\infty} a^k$, shows that (x_n) is a Cauchy sequence and therefore (as X is complete) that $x_n \to x$ for some $x \in X$. From (2.2.9) we then have

$$x = \lim_{n\to\infty} x_n = \lim_{n\to\infty} F(x_{n-1}) = F(x), \tag{2.2.13}$$

where for the last equality we have used Theorem 2.2.2. Finally, to show the uniqueness of the fixed point x, assume by way of contradiction that there is an $y \in X$ with $y \ne x$ such that $F(y) = y$. Then by (2.2.8) we have

$$d(x,y) = d(F(x), F(y)) \le a d(x,y) < d(x,y),$$

which is absurd. This completes the proof of Theorem 2.2.4.

More on metric spaces

Closed sets. Closure of a set

Let (X, d) be a metric space.

Definition 2.2.11. A subset $A \subset X$ is said to be *closed* if its complement $A^c = X \setminus A$ is open.

It follows that X itself and the empty set $\emptyset$ are closed, and so is a finite union of closed sets and an arbitrary intersection of closed sets. The interval $[a, b] = \{x \in \mathbb{R} : a \leq x \leq b\}$ is a closed subset of $\mathbb{R}$, for $\mathbb{R} \setminus [a, b] =]-\infty, a[\cup]b, +\infty[$. More generally, consider the following.

Exercise 2.2.10. Show that the closed ball $B'(x, r) = \{y \in X : d(y, x) \leq r\}$ is a closed set.

Proposition 2.2.1. *In a metric space, a subset A is closed if and only if for any convergent sequence* $(x_n) \subset A$ *we have* $\lim x_n \in A$.

Proof. Assume first that A is closed and suppose by way of contradiction that there exists a sequence $(x_n) \subset A$ which is convergent but such that $x \equiv \lim x_n \notin A$. Then $x \in A^c$, and as A^c is open it follows that $B(x, r) \subset A^c$ for some $r > 0$. As $x_n \to x$, we also have $x_n \in B(x, r)$ for n sufficiently large, thus contradicting the fact that $(x_n) \subset A$.

Suppose now that for any convergent sequence $(x_n) \subset A$ we have $\lim x_n \in A$. We claim that A is then necessarily closed, for otherwise A^c would not be open, and therefore there would be a point $z \in A^c$ not interior to A^c. Thus, for any $r > 0$, $B(z, r) \not\subset A^c$, that is, $B(z, r) \cap A \neq \emptyset$. Therefore, taking $r = \frac{1}{n}$ $(n \in \mathbb{N})$ we obtain in particular

$$\forall n \in \mathbb{N} \quad \exists x_n \in A : \quad d(x_n, z) < \frac{1}{n}.$$

We have thus constructed a sequence $(x_n) \subset A$ which is convergent, but whose limit $z \notin A$, contradicting our assumption. □

Definition 2.2.12. Given a subset A of X, a point $x \in X$ is said to be a *cluster point* of A if any neighborhood of x contains points of A, that is, if

$$B(x, r) \cap A \neq \emptyset \quad \text{for any } r > 0.$$

The set of all cluster points of A is called the *closure* of A and is denoted with $\overline{A}$.

Example 2.2.9. Let $A \subset \mathbb{R}$ be bounded from above; then $L = \sup A$ is a cluster point of A (which may or may not belong to A; in the former case, it is the maximum of A). Similarly, if A is bounded from below, then $\inf A \in \overline{A}$.

Evidently, we have $A \subset \overline{A}$. Further properties of $\overline{A}$ can be gained through the following exercises.

Exercise 2.2.11. For any set $A \subset X$, we have $(\overline{A})^c = \text{int}(A^c)$.

Using the corresponding statements concerning open sets and the interior of a set (see in particular Exercise 2.2.6), we then have the following proposition.

Proposition 2.2.2. *The closure $\overline{A}$ of a set A is a closed set. A is closed if and only if $A = \overline{A}$.*

Exercise 2.2.12. If $A \subset B$, then $\overline{A} \subset \overline{B}$.

Exercise 2.2.13. It follows in particular from Exercises 2.2.10 and 2.2.12 and Proposition 2.2.2 that $\overline{B(x,r)} \subset B'(x,r)$. Give an example in which strict inclusion holds and show that on the other hand $\overline{B(x,r)} = B'(x,r)$ if $X = \mathbb{R}^n$.

Finally, arguing as in Proposition 2.2.1, we easily get a characterization of $\overline{A}$ via convergent sequences.

Proposition 2.2.3. *Let $A \subset X$ and let $x \in X$. Then $x \in \overline{A}$ if and only if there exists a sequence $(x_n) \subset A$ such that* $\lim x_n = x$.

Bounded sets. Diameter of a set

Recall that a subset $A \subset \mathbb{R}$ is said to be bounded if it is bounded from above and from below, that is, if there exist $m, M \in \mathbb{R}$ such that $m \le x \le M$ for all $x \in A$. Taking $R = \max\{|m|, |M|\}$, we then easily verify that $|x| \le R$ for all $x \in A$ and in turn that $|x - y| \le 2R$ for any $x, y \in A$.

Definition 2.2.13. A subset A of a metric space X with distance d is said to be *bounded* if there exists $k > 0$ such that $d(x,y) \le k$ for all $x, y \in A$. In this case the number $d(A) \equiv \sup\{d(x,y) : x, y \in A\}$ is called the *diameter* of A.

For instance, $d(B(x,r)) \le 2r$. The expected equality $d(B(x,r)) = 2r$ holds if $X = \mathbb{R}^n$ or, more generally, if X is a normed vector space.

Proposition 2.2.4. *Let $A \subset X$. Then the following properties are equivalent:*
(i) *$A \subset B(x,R)$ for some $x \in X$ and some $R > 0$;*
(ii) *A is bounded;*
(iii) *for any $x \in X$, there exists an $R > 0$ such that $A \subset B(x,R)$.*

Proof. This is left as an exercise.

A sequence $(x_n) \subset X$ is said to be bounded if the set of its points $\{x_n : n \in \mathbb{N}\}$ is a bounded subset of X. More generally, a function f of a set S into a metric space X is said to be bounded if the image $f(S) = \{f(x) : x \in S\}$ is a bounded subset of X. □

Exercise 2.2.14. Any convergent sequence is bounded.

Exercise 2.2.15. If a subset V of X is unbounded, then there exists a sequence $(x_n) \subset V$ such that $d(x_n, z) \to \infty$, where z is any chosen point of X. (Use the fact that the inclusion $V \subset B(z,n)$ is false for every $n \in \mathbb{N}$.)

Exercise 2.2.16. Any bounded closed subset A of $\mathbb{R}$ has a minimum and maximum; indeed, $\sup A$ and $\inf A$ are cluster points of A (Example 2.2.9), and therefore by Proposition 2.2.2 they belong to A because A is closed.

2.3 Some function spaces

Let A be any set. We denote with $B(A)$ the set of all bounded real-valued functions defined on A:

$$B(A) = \{f : A \to \mathbb{R} \mid f \text{ is bounded}\}.$$

For $f, g \in B(A)$ put

$$d(f, g) = \sup_{x \in A} |f(x) - g(x)|. \tag{2.3.1}$$

Then $d(f, g)$ is well defined and indeed we have the following proposition.

Proposition 2.3.1. *The mapping d defined by* (2.3.1) *on $B(A) \times B(A)$ is a distance on $B(A)$.*

The proof of Proposition 2.3.1 (that is, the verification of properties (i), (ii), and (iii) of Definition 2.2.1) follows by the properties (2.1.1) and (2.1.2) of the lowest upper bound and by the properties of the absolute value $|.|$ in $\mathbb{R}$, and is left as an exercise.

The significance of the metric d introduced in (2.3.1) becomes clear if we go back to the definition of *uniform convergence* (Definition 2.1.1).

Proposition 2.3.2. *Let (f_n) be a sequence in $B(A)$ and let $f \in B(A)$. Then $f_n \to f$ in the metric space $(B(A), d)$ if and only if $f_n \to f$ uniformly on A.*

The implications of the uniform convergence rest, to a good extent, on the following two theorems.

Theorem 2.3.1. *The metric space $(B(A), d)$, with d given by* (2.3.1), *is complete.*

Proof. Let (f_n) be a Cauchy sequence in $B(A)$; we need to prove that there exists $f \in B(A)$ such that $f_n \to f$ in $B(A)$.

(a) By (2.3.1), we have

$$\forall x \in A, \quad \forall n, m \in \mathbb{N} \quad |f_n(x) - f_m(x)| \le d(f_n, f_m). \tag{2.3.2}$$

Hence, for any $x \in A$, $(f_n(x))$ is a Cauchy sequence in $\mathbb{R}$, and therefore it converges in $\mathbb{R}$. We can thus define a function $f : A \to \mathbb{R}$ by the rule

$$f(x) \equiv \lim_{n \to \infty} f_n(x) \quad (x \in A); \tag{2.3.3}$$

f is precisely the *pointwise limit* of the sequence (f_n) (Section 2.1).

(b) We now claim that $f \in B(A)$ and that $d(f_n, f) \to 0$. To this purpose fix $\epsilon > 0$ and let $n_0 \in \mathbb{N}$ be such that $d(f_n, f_m) < \epsilon$ for any $n, m \in \mathbb{N}$ with $n, m \geq n_0$. Then by (2.3.2),

$$\forall x \in A, \quad \forall n, m \geq n_0 \quad |f_n(x) - f_m(x)| < \epsilon, \tag{2.3.4}$$

whence, letting $m \to \infty$ and using the definition (2.3.3) of f, we have

$$\forall x \in A, \quad \forall n \geq n_0 \quad |f_n(x) - f(x)| \leq \epsilon. \tag{2.3.5}$$

This first shows that f is bounded, for

$$|f(x)| \leq |f(x) - f_{n_0}(x)| + |f_{n_0}(x)| \leq \epsilon + K_0$$

for all $x \in A$ and some $K_0 \in \mathbb{R}$. Moreover, (2.3.5) also yields

$$d(f_n, f) = \sup_{x \in A} |f_n(x) - f(x)| \leq \epsilon$$

for all $n \geq n_0$, showing that (f_n) converges to f in $B(A)$. □

Suppose now that $A \subset \mathbb{R}$ and set

$$C_B(A) = C(A) \cap B(A),$$

where $C(A) = \{f : A \to \mathbb{R} | f \text{ is continuous}\}$. We then have the following theorem.

Theorem 2.3.2. *The metric space $(C_B(A), d)$, with d given by* (2.3.1), *is complete.*

Proof. Let (f_n) be a Cauchy sequence in $C_B(A)$; then it is a Cauchy sequence in $B(A)$ and therefore, by Theorem 2.3.1, it converges to an $f \in B(A)$ in the metric (2.3.1). It remains to show that f is continuous in A. To this purpose, fix a point $x_0 \in A$ and an $\epsilon > 0$. Then because of the uniform convergence of f_n to f, there exists an index $n_0 \in \mathbb{N}$ such that $d(f_n, f) < \epsilon/3$ for any $n \geq n_0$. On the other hand, as each f_n is continuous by assumption, there also exists a $\delta (= \delta_{n_0}) > 0$ such that $|f_{n_0}(x) - f_{n_0}(x_0)| < \epsilon/3$ whenever $x \in A$ and $|x - x_0| < \delta$. It follows that for such points x we have

$$|f(x) - f(x_0)| \leq |f(x) - f_{n_0}(x)| + |f_{n_0}(x) - f_{n_0}(x_0)| + |f_{n_0}(x_0) - f(x_0)| < \epsilon,$$

which shows the continuity of f at x_0 and thus ends the proof of Theorem 2.3.2. □

It is convenient to state, independently from completeness, the basic relation between uniform convergence and continuity that comes to light in the proof of Theorem 2.3.2:

Proposition 2.3.3. *If a sequence of bounded continuous functions from A to $\mathbb{R}$ converges **uniformly** on A to a bounded function f, then f is continuous.*

Remark 2.3.1. Theorem 2.3.2 and Proposition 2.3.3 hold unaltered if A is a subset of $\mathbb{R}^n$ (or in fact of any metric space): in the proof, we just have to replace the interval $]x_0 - \delta, x_0 + \delta[$ with the ball $B(x_0, \delta)$. Now if we take in particular $A \subset \mathbb{R}^n$ to be closed and bounded, then we can invoke the Weierstrass theorem, Theorem 0.0.3, to ensure that any continuous function defined on A is bounded and in fact attains its maximum and minimum value; therefore, in such case we have

$$C_B(A) = C(A)$$

and the distance (2.3.1) becomes

$$d(f,g) = \max_{x \in A} |f(x) - g(x)|. \tag{2.3.6}$$

We thus have the following result.

Corollary 2.3.1. *If $K \subset \mathbb{R}^n$ is closed and bounded, then $C(K)$ – equipped with the distance* (2.3.6) *– is a complete metric space.*

Application: local existence and uniqueness of solutions to the IVP for first-order ODEs

With the help of the previous two sections, we are ready to state and prove one of the simplest existence and uniqueness results for solutions of the IVP (Chapter 1, Section 1.1)

$$\begin{cases} x' = f(t,x) \\ x(t_0) = x_0, \end{cases} \tag{2.3.7}$$

where f is a real function defined on an open set $A \subset \mathbb{R}^2$ and $(t_0, x_0) \in A$. Recall that f is said to be *Lipschitzian with respect to the second variable in A* if there exists a constant $L > 0$ such that

$$|f(t,x) - f(t,y)| \le L|x - y| \quad \forall (t,x),(t,y) \in A. \tag{2.3.8}$$

We consider the particular case in which A is an (open) **strip** in the plane: this geometrical assumption helps to make transparent the proof of the statement below.

Theorem 2.3.3. *Let $f : I \times \mathbb{R} \to \mathbb{R}$, with I an open interval in $\mathbb{R}$. Assume that:*

(a) *f is continuous in $I \times \mathbb{R}$;*
(b) *f is Lipschitzian with respect to the second variable in $I \times \mathbb{R}$.*

Then given any $(t_0, x_0) \in I \times \mathbb{R}$, there exists a neighborhood I_0 of t_0, $I_0 \subset I$, such that (2.3.7) *has a* unique *solution defined in I_0.*

Proof. Fix $(t_0, x_0) \in I \times \mathbb{R}$ and let $r_1 > 0$ be such that $]t_0 - r_1, t_0 + r_1[\subset I$. Then let r_0 be such that $0 < r_0 < \min\{r_1, 1/L\}$, where L is as in (2.3.8), and put $I_0 = [t_0 - r_0, t_0 + r_0]$. By

Lemma 1.1.2, proving the theorem is equivalent to proving the existence and uniqueness of a fixed point for the map $F : C(I_0) \to C(I_0)$ defined in (1.1.7) and relative to I_0, that is,

$$F(u)(t) = x_0 + \int_{t_0}^{t} f(s, u(s))\, ds, \quad t \in I_0. \tag{2.3.9}$$

However, by Corollary 2.3.1, $X = C(I_0)$ – equipped with the distance (2.3.6) relative to I_0 – is a complete metric space and thus by the contraction mapping theorem, it will be enough to show that

$$d(F(u), F(v)) \leq \alpha d(u, v) \tag{2.3.10}$$

for some α with $0 < \alpha < 1$ and for all $u, v \in X$. Indeed, this follows from the inequalities (holding for all $t \in I_0$ with $t > t_0$)

$$\begin{aligned} |F(u)(t) - F(v)(t)| &\leq \int_{t_0}^{t} |f(s, u(s)) - f(s, v(s))|\, ds \\ &\leq \int_{t_0}^{t} L|u(s) - v(s)|\, ds \leq Ld(u, v)|t - t_0| \leq Lr_0 d(u, v). \end{aligned} \tag{2.3.11}$$

Arguing similarly for the case $t < t_0$ and taking the sup of the first term in this chain of inequalities, we obtain (2.3.10) with $\alpha = Lr_0 < 1$, whence the conclusion follows. □

Theorem 2.3.3 can be greatly improved to give the standard form of the local existence and uniqueness principle for solutions of (2.3.7), which requires merely that f be *locally* Lipschitzian with respect to the second variable and puts no restriction on the open set A where f is defined; see the statement given in Chapter 1, Theorem 1.1.1, and see for instance Hale [5], Walter [1], and Dieudonné [7] for proofs of it.

2.4 Banach spaces

Definition 2.4.1. A *Banach space* is a normed vector space E that is *complete* as a metric space with the distance induced by the norm (see (2.2.2)),

$$d(x, y) \equiv \|x - y\| \quad (x, y \in E). \tag{2.4.1}$$

Example 2.4.1. $\mathbb{R}^n$ endowed with the Euclidean norm $\|x\| = (\sum_1^n x_i^2)^{\frac{1}{2}}$ (or with any other norm, see Exercise 2.7.8) is a Banach space.

Example 2.4.2. The function spaces $B(A)$, $C_B(A)$ considered in Section 3.3 equipped with the norm

$$\|f\| = \sup_{x\in A}|f(x)| \tag{2.4.2}$$

inducing the distance (2.3.1) of the uniform convergence in A are Banach spaces, as shown in Theorems 2.3.1 and 2.3.2.

The latter example can be generalized in two directions. The first is about the domain and range of the functions that one needs to deal with, for we can consider the space

$$B(A;\mathbb{R}^m) = \{f : A \to \mathbb{R}^m \mid f \text{ is bounded}\} \tag{2.4.3}$$

with A any set and the space

$$C_B(A;\mathbb{R}^m) = \{f : A \to \mathbb{R}^m \mid f \text{ is bounded and continuous}\} \tag{2.4.4}$$

with $A \subset \mathbb{R}^n$. Equipped with the norm

$$\|f\| = \sup_{x\in A}\|f(x)\|,$$

these are Banach spaces by virtue of the completeness of the target space $\mathbb{R}^m$, as becomes clear by inspecting the proof of Theorems 2.3.1 and 2.3.2. Actually, the same arguments yield – more generally – the completeness of the spaces

$$B(A;E) \quad \text{and} \quad C_B(A;E),$$

which are defined in complete analogy with (2.4.3) and (2.4.4), respectively, replacing the target space $\mathbb{R}^m$ with a general Banach space E. In the same vein, one might consider $C_B(A;E)$ when A is a subset of any metric space, and so on, if useful or necessary.

The second line to follow in order to obtain new Banach spaces from classes of functions concerns their *regularity*, meaning their property of being differentiable up to a given order. Consider first the space

$$C^1([a,b]) \equiv \{f : [a,b] \to \mathbb{R} : f \text{ is differentiable in } [a,b] \text{ and } f' \text{ is continuous in } [a,b]\}. \tag{2.4.5}$$

It is easily checked that this is a vector space and that

$$\|f\|_1 \equiv \|f\| + \|f'\| \tag{2.4.6}$$

is a norm on it. We ask whether this is a Banach space, which we know to be true for $C([a,b])$. The affirmative answer to this question rests on Theorem 2.1.3 and its Corollary 2.1.1. Indeed, let $(f_n) \subset C^1([a,b])$ be a Cauchy sequence. By (2.4.6), we see that both (f_n) and (f_n') are Cauchy sequences in $C([a,b])$. As this is complete, we thus find two continuous functions f, g on $[a,b]$ such that

$$f_n \to f, \quad f_n' \to g \quad \text{in} \quad C([a,b]),$$

that is, uniformly on $[a,b]$. However, Corollary 2.1.1 then ensures that $f \in C^1(I)$ and $f' = g$; using this and (2.4.6), we conclude that

$$\|f_n - f\|_1 \equiv \|f_n - f\| + \|f_n' - f'\| \to 0 \tag{2.4.7}$$

as $n \to \infty$, proving that (f_n) converges to f in $C^1([a,b])$ and thus proving the completeness of the latter. In the same way one proves that given any $k \in \mathbb{N}$, the space of k-times continuously differentiable functions on $[a,b]$

$$C^k([a,b]) \equiv \{f : [a,b] \to \mathbb{R} : f \text{ is } k \text{ times diff. in } [a,b] \text{ and } f_i' \in C([a,b])\ \forall i = 1,\ldots,k\} \tag{2.4.8}$$

is a Banach space for the norm

$$\|f\|_k \equiv \|f\| + \|f'\| + \cdots + \|f_k'\|. \tag{2.4.9}$$

The spaces $C^k(\overline{\Omega})$, $\Omega \subset \mathbb{R}^n$

The discussion provided so far for continuous or regular (real-valued) functions of one real variable can be naturally extended to functions of several real variables, replacing the open bounded interval $]a,b[\subset \mathbb{R}$ with an open bounded subset $\Omega \subset \mathbb{R}^n$. Thus, starting with the space

$$C(\overline{\Omega}) = \{f : \overline{\Omega} \to \mathbb{R} \mid f \text{ continuous}\}, \tag{2.4.10}$$

which is a Banach space for the norm (2.4.2), we will also meet the spaces $C^k(\overline{\Omega})$ $(k \geq 1)$ of k-times continuously differentiable functions on $\overline{\Omega}$; each of them is a Banach space when equipped with the norm similar to (2.4.9) and involving all partial derivatives up to and including those of order k of f. For instance, the norm in $C^1(\overline{\Omega})$ is given by

$$\|f\|_{C^1(\overline{\Omega})} \equiv \|f\| + \sum_{i=1}^{n} \left\| \frac{\partial f}{\partial x_i} \right\|. \tag{2.4.11}$$

Note that the partial derivatives at points of the boundary $\partial\Omega$ of Ω must be defined appropriately. See Section 4.1 of Chapter 4 for some remarks on this issue.

Convergence of series in normed and Banach spaces

In a vector space E, we can by definition make the sum of two, and therefore of a finite number of, vectors of E. It is important (both conceptually and practically) to give meaning to “infinite sums” of vectors. If we have a norm in E, this can be plainly done in complete analogy with the limit process that we use for series of real numbers.

Definition 2.4.2. Let E be a normed vector space with norm $\|.\|$. If (x_n) is a sequence of vectors of E, we say that *the series* $\sum_{n=1}^{\infty} x_n$ *converges* if the sequence (s_n) of the *partial sums* of the series, defined putting

$$s_n = \sum_{i=1}^{n} x_i \quad (n \in \mathbb{N}),$$

converges in E; that is, if there exists an $s \in E$ such that $s_n \to s$, i. e., $\|s_n - s\| \to 0$, as $n \to \infty$. In this case we write

$$s = \sum_{n=1}^{\infty} x_n \tag{2.4.12}$$

and call s the *sum* of the series $\sum_{n=1}^{\infty} x_n$.

Theorem 2.4.1. *Let E be a Banach space and let $(x_n) \subset E$. If the numerical series $\sum_{n=1}^{\infty} \|x_n\|$ converges, then the series $\sum_{n=1}^{\infty} x_n$ converges in E.*

Proof. Let (s_n), $s_n = \sum_{i=1}^{n} x_i$ $(n \in \mathbb{N})$, be the sequence of the partial sums of the series $\sum_{n=1}^{\infty} x_n$. For any $n, m \in \mathbb{N}$ we have, assuming for instance that $n > m$,

$$\|s_n - s_m\| = \left\| \sum_{i=1}^{n} x_i - \sum_{i=1}^{m} x_i \right\| = \left\| \sum_{i=m+1}^{n} x_i \right\| \leq \sum_{i=m+1}^{n} \|x_i\|,$$

so that, putting $t_n = \sum_{i=1}^{n} \|x_i\|$, we conclude that

$$\|s_n - s_m\| \leq \sum_{i=m+1}^{n} \|x_i\| = |t_n - t_m| \quad \forall n, m \in \mathbb{N}. \tag{2.4.13}$$

Our assumption implies that (t_n) is a convergent sequence, and (2.4.13) thus implies that $(s_n) \subset E$ is a Cauchy sequence and therefore converges in E as E is complete by assumption. □

We have seen a first application of Theorem 2.4.1 in Section 1.4 of Chapter 1, dealing with the definition of the exponential matrix e^A of a given matrix A, and showing its relevance for first-order systems of linear ODEs with constant coefficients. A second fundamental consequence is discussed next.

Total convergence of a series of bounded functions

Suppose that (f_n) is a sequence of bounded real functions defined in a set A and suppose that the numerical series

$$\sum_{n=1}^{\infty} \|f_n\|, \quad \|f_n\| = \sup_{x \in A} |f_n(x)| \tag{2.4.14}$$

is convergent, in which case the series $\sum_{n=1}^{\infty} f_n$ is said to be *totally convergent*. Then $\sum_{n=1}^{\infty} f_n$ converges uniformly in A; this follows immediately on applying Theorem 2.4.1 to the Banach space $E = B(A)$ recalled in Example 2.4.2. An equivalent way of saying that the total convergence implies uniform convergence is of course given by the *Weierstrass M-test*, stated without proof in Proposition 2.1.1. It also follows from Proposition 2.3.3 (applied to the sequence (s_n) of the partial sums of the series $\sum_{n=1}^{\infty} f_n$) that if each function f_n of a totally convergent series is continuous, then so is the sum f of the series; moreover, the series can be integrated term-by-term as stated in Corollary 2.1.2. Likewise, on the basis of Corollary 2.1.3, the property that $f \in C^1(I)$ if each term $f_n \in C^1(I)$ is granted if the two series

$$\sum_{n=1}^{\infty} \|f_n\|, \ \sum_{n=1}^{\infty} \|f_n'\|$$

are both convergent.

2.5 Compactness

Subsequences

Given a sequence $(x_n)_{n\in\mathbb{N}}$ of points of a set X and a strictly increasing mapping $k \to n_k$ of $\mathbb{N}$ into itself, we say that the sequence $(x_{n_k})_{k\in\mathbb{N}}$ is a *subsequence* of $(x_n)_{n\in\mathbb{N}}$. For example, $(\frac{1}{k^2})$ is a subsequence of $(\frac{1}{n})$; likewise, (x_{k^3}), (x_{2k}), (x_{2k-1}) – and of course (x_k) itself – are subsequences of the given sequence (x_n).

Proposition 2.5.1. *Let (X, d) be a metric space, let $(x_n) \subset X$, and let $x_0 \in X$. If $x_n \to x_0$ as $n \to \infty$, then $x_{n_k} \to x_0$ as $k \to \infty$ for* any *subsequence (x_{n_k}) of (x_n).*

The proof of Proposition 2.5.1 follows immediately from the definitions and the inequality $n_k \geq k (k \in \mathbb{N})$, which is easily proven to hold for any strictly increasing mapping $k \to n_k$ of $\mathbb{N}$ into $\mathbb{N}$.

The Bolzano–Weierstrass theorem

Theorem 2.5.1. *Any sequence of points in a closed, bounded interval of the real line contains a convergent subsequence.*

Proof. Let $[a, b] \subset R$ be as in the statement and let (x_n) be any sequence with values in $[a, b]$. We first show that there exists a sequence $(T_k)_{k\in\mathbb{N}}$ of closed subintervals of $[a, b]$,

$$T_k = [a_k, b_k] \subset [a, b] \equiv T_0$$

having the following properties:

(i) $T_k \subset T_{k-1}$;

(ii) $b_k - a_k = \frac{b-a}{2^k}$;

(iii) the set of indices $I_k \equiv \{n \in \mathbb{N} : x_n \in T_k\}$ is infinite.

To construct the sequence (T_k), start cutting $[a,b] = T_0$ in its midpoint obtaining two intervals of equal length $\frac{b-a}{2}$; at least one of them contains the points x_n for infinitely many indices $n \in \mathbb{N}$; choose it and call it T_1, and then iterate the procedure. We now construct a subsequence $(x_{n_k})_{k\in\mathbb{N}}$ of (x_n) as follows. Let $n_1 = \min I_1$ and for each integer $k > 1$ put

$$n_k = \min\{n \in I_k : n > n_{k-1}\} = \min\{n \in \mathbb{N} : x_n \in T_k \text{ and } n > n_{k-1}\}.$$

In particular, $x_{n_k} \in T_k$ for each $k \in \mathbb{N}$. We claim that the sequence $(x_{n_k})_{k\in\mathbb{N}}$ has the following property: for any $k_0 \in \mathbb{N}$ and any $k, h \in \mathbb{N}$ with $k, h \geq k_0$ we have

$$|x_{n_k} - x_{n_h}| \leq \frac{b-a}{2^{k_0}}. \tag{2.5.1}$$

Indeed, if $k \geq k_0$, then by (i) $x_{n_k} \in T_{k_0}$, and similarly for x_{n_h}; therefore, $|x_{n_k} - x_{n_h}| \leq b_{k_0} - a_{k_0}$, and thus (ii) yields (2.5.1). The latter inequality shows that (x_{n_k}) is a Cauchy sequence of real numbers, and the completeness of $\mathbb{R}$ proves that (x_{n_k}) converges to some $x_0 \in \mathbb{R}$; however, as $[a,b]$ is a closed subset of $\mathbb{R}$, it follows that $x_0 \in [a,b]$. This ends the proof of Theorem 2.5.1. □

Compact sets

Definition 2.5.1. A subset K of a metric space X is said to be *compact* if any sequence $(x_n) \subset K$ contains a subsequence converging to a point of K.

The whole space X may be compact; we then speak of a compact metric space. Theorem 2.5.1 can then be rephrased as follows.

Theorem 2.5.2. *Any closed, bounded interval $[a,b]$ is a compact subset of $\mathbb{R}$.*

The next two statements show some basic properties of compact sets.

Theorem 2.5.3. *Let X be a metric space and let $K \subset X$. Then:*
(a) *if K is compact, then K is closed and bounded;*
(b) *if $F \subset X$ is compact, $K \subset F$, and K is closed, then K is compact.*

Proof. This is left as an exercise.

Let us now go back to the special case $X = \mathbb{R}^n$, in which – by virtue of the Bolzano–Weierstrass theorem – one can *characterize* the compact sets. An n-dimensional (closed and bounded) interval is by definition a subset $I \subset \mathbb{R}^n$ of the form

$$I = [a_1, b_1] \times [a_2, b_2] \times \cdots \times [a_n, b_n].$$ □

Proposition 2.5.2. *Any n-dimensional interval is a compact subset of $\mathbb{R}^n$.*

Proof. It is enough to deal with the case $n = 2$. Let (z_n) be any sequence in $I = [a,b] \times [c,d]$. Put $z_n = (x_n, y_n)$ and let (x_{n_k}) be a subsequence of (x_n) that converges to a point

$x_0 \in [a,b]$ (Theorem 2.5.2). Consider the corresponding subsequence (y_{n_k}) of (y_n); again by Theorem 2.5.2, this will contain a further subsequence $(y_{n_{k_i}})$ converging to some $y_0 \in [c,d]$. Now the subsequence $(z_{n_{k_i}})$ of (z_n),

$$z_{n_{k_i}} = (x_{n_{k_i}}, y_{n_{k_i}}),$$

will converge to $z_0 \equiv (x_0, y_0) \in I$, and this proves our claim. □

Theorem 2.5.4. *Any closed and bounded subset of $\mathbb{R}^n$ is compact.*

Proof. If $K \subset \mathbb{R}^n$ is bounded, we can find an n-dimensional interval I such that $K \subset I$. As I is compact by Proposition 2.5.2, it follows by part (b) of Theorem 2.5.3 that K too is compact if it is further assumed to be closed, as we do.

Using also part (a) of Theorem 2.5.3, we therefore conclude that a subset of $\mathbb{R}^n$ is compact *if and only if* it is closed and bounded. □

Compactness and continuity

Theorem 2.5.5. *Let X, Y be metric spaces, let $K \subset X$, and let $f : K \to Y$. If f is continuous and K is compact, then $f(K)$ is a compact subset of Y.*

Proof. Let (y_n) be a sequence in $f(K)$, and for each $n \in \mathbb{N}$ let $x_n \in K$ be such that $f(x_n) = y_n$. As K is compact, the sequence (x_n) contains a subsequence (x_{n_k}) converging to some $x_0 \in K$. Using the continuity of f and Theorem 2.2.2, it follows that $f(x_{n_k}) \to f(x_0)$ as $k \to \infty$. That is to say, the subsequence (y_{n_k}) of (y_n) converges to $f(x_0) \in f(K)$, as desired. □

Theorem 2.5.6. *Let X be a metric space, let $K \subset X$, and let $f : K \to \mathbb{R}$. If f is continuous and K is compact, then f attains its minimum and its maximum value in K; that is, there exist $x_1, x_2 \in K$ such that*

$$f(x_1) \le f(x) \le f(x_2) \quad \forall x \in K. \tag{2.5.2}$$

Proof. Indeed, it follows by Theorem 2.5.5 and by (a) of Theorem 2.5.3 that $f(K)$ is a closed bounded subset of $\mathbb{R}$. Therefore – see Exercise 2.2.16 – there exist $y_1, y_2 \in f(K)$ such that $y_1 \le y \le y_2$ for all $y \in f(K)$, whence (2.5.2) follows on putting $y_i = f(x_i)$ with $x_i \in K$ $(i = 1, 2)$.

As a special case, we obtain the familiar Weierstrass theorem recalled in the Preliminaries (Theorem 0.0.3). □

Theorem 2.5.7. *Let $K \subset \mathbb{R}^n$ and let $f : K \to \mathbb{R}$. If f is continuous and K is closed and bounded, then f attains its minimum and its maximum value in K.*

Example 2.5.1 (Not every closed bounded set is compact). Let $X = C([0,1])$ with the norm $\|x\| = \max_{0 \le t \le 1} |x(t)|$. The closed ball $B'(0,1) = \{x \in X : \|x\| \le 1\}$ is closed and bounded, but *not* compact. To see this, consider for instance the sequence $(x_n) \subset$

$K \equiv B'(0,1)$ defined putting $x_n(t) = t^n$ for $0 \le t \le 1$ and $n \in \mathbb{N}$. Then (x_n) contains no convergent subsequence, for if it did, calling (x_{n_k}) this subsequence and $x_0 \in X$ its limit, we should have in particular

$$\lim_{k\to\infty} x_{n_k}(t) = x_0(t) \quad \forall t \in [0,1],$$

which leads to a contradiction, since x_0 should be continuous on $[0,1]$ while $x_0(t) = 0$ for $0 \le t < 1$ and $x_0(1) = 1$.

Evidently, any bounded sequence (x_n) in $C([a,b])$ converging pointwisely to a discontinuous function would equally well serve as a counterexample to the statement "K closed and bounded $\Rightarrow K$ compact."

Compactness and uniform continuity (Heine–Cantor's theorem)

Theorem 2.5.8. *Let X, Y be metric spaces, let $K \subset X$, and let $f : K \to Y$. If f is continuous and K is compact, then f is* uniformly *continuous.*

Proof. Suppose by way of contradiction that f is **not** uniformly continuous, so that (by Definition 2.2.8) for some $\epsilon_0 > 0$, it happens that

$$\forall \delta > 0, \quad \exists x, y \in K : d(x,y) < \delta \quad \text{and} \quad d'(f(x), f(y)) \ge \epsilon_0.$$

Choosing $\delta = \frac{1}{n} (n \in \mathbb{N})$, we find therefore two sequences $(x_n), (y_n) \subset K$ such that

$$d(x_n, y_n) < \frac{1}{n} \quad \text{and} \quad d'(f(x_n), f(y_n)) \ge \epsilon_0, \quad \forall n \in \mathbb{N}. \tag{2.5.3}$$

Let (x_{n_k}) be a subsequence of (x_n) converging to a point $x_0 \in K$. As $d(x_{n_k}, y_{n_k}) \le \frac{1}{n_k}$ and $n_k \ge k$ for all $k \in \mathbb{N}$, it follows that also the subsequence (y_{n_k}) of (y_n) converges to x_0. Therefore, again by Theorem 2.2.2, it follows that, as $k \to \infty$,

$$f(x_{n_k}) \to f(x_0) \quad \text{and} \quad f(y_{n_k}) \to f(x_0).$$

In turn, this implies by the triangle inequality that $d'(f(x_{n_k}), f(y_{n_k})) \to 0$ as $k \to \infty$. However, this contradicts the second inequality in (2.5.3) and thus proves the theorem. □

Example 2.5.2. When the domain K of f is not compact, f may be continuous but not uniformly continuous, as the following example shows. Let $X = \mathbb{R}$, let $K =]0,1]$, and let $f(x) = \frac{1}{x}$ for $x \in K$. Then f is not uniformly continuous in K, because for any $\delta > 0$ we can find $x, y \in K$ with $|x - y| < \delta$ but $|f(x) - f(y)| \ge 1$. To see this it is enough, given any $\delta > 0$, to take x with $0 < x < \min\{\delta, 1\}$ and $y = \frac{x}{2}$; we then have

$$|x - y| = \frac{x}{2} < x < \delta \quad \text{and} \quad \left|\frac{1}{x} - \frac{1}{y}\right| = \left|\frac{1}{x} - \frac{2}{x}\right| = \frac{1}{x} > 1.$$

Example 2.5.3 (Continuity of the Nemytskii operator). Let $X = C([a,b])$ equipped with the norm $\|x\| = \max_{a\le t\le b}|x(t)|$, let $f : [a,b]\times\mathbb{R}\to\mathbb{R}$ be continuous, and let $N_f : X \to X$ be the Nemytskii operator induced by f (Remark 1.1.1). Then N_f is continuous. To prove this, we can invoke once more Theorem 2.2.2 and show that for any sequence $(x_n)\subset X$ convergent to some x_0, $(N_f(x_n))$ converges to $N_f(x_0)$ in X. To this purpose, fix $\epsilon > 0$; we need to find an $n_0 \in \mathbb{N}$ such that for every $n \ge n_0$ we have

$$|N_f(x_n)(t) - N_f(x_0)(t)| = |f(t,x_n(t)) - f(t,x_0(t))| < \epsilon \quad \forall t \in [a,b]. \tag{2.5.4}$$

Fix $R > 0$ and let $K = [-R,R]$. By Heine–Cantor's theorem, f is *uniformly* continuous in $[a,b]\times K$, and thus there is a $\delta > 0$ such that for any $(t,x),(s,y) \in [a,b]\times K$ with $|t-s| + |x-y| < \delta$, we have $|f(t,x) - f(s,y)| < \epsilon$. In particular, for any $x,y \in K$ with $|x-y| < \delta$ we have

$$|f(t,x) - f(t,y)| < \epsilon \quad \forall t \in [a,b]. \tag{2.5.5}$$

We can assume that $\delta \le R$, and as (x_n) converges uniformly to x_0, we can find an $n_0 \in \mathbb{N}$ such that, for every $n \ge n_0$, the inequality

$$|x_n(t) - x_0(t)| < \delta \tag{2.5.6}$$

is satisfied for *every* $t \in [a,b]$; using this in (2.5.5), we obtain (2.5.4) and thus prove that $(N_f(x_n))$ converges to $N_f(x_0)$ uniformly on $[a,b]$, as desired.

2.6 Connectedness

Definition 2.6.1. A metric space X is said to be *disconnected* if there exist two subsets A, B of X such that:
(i) $A \ne \emptyset, B \ne \emptyset$;
(ii) $A \cap B = \emptyset, A \cup B = X$;
(iii) both A and B are open.

In other words, X is disconnected if it is the union of two open, non-empty, disjoint subsets of X.

Remark 2.6.1. It follows immediately from this definition that X is disconnected if and only if it has a subset A with $A \ne \emptyset$, $A \ne X$, and A is both open and closed in X.

Of course Definition 2.6.1 applies also to a subset F of X, meaning that the metric space (F, d_F) is disconnected, with d_F being the restriction to $F \times F$ of the distance d given on X. This amounts to requiring that F is the disjoint union of two non-empty subsets A and B of F which are open *in* (F, d_F), or – as is usually said – open *relative to F*. There is an easy characterization of the relatively open subsets of a given set F in X.

Proposition 2.6.1. *Let $F \subset X$. A subset A of F is open relative to F iff there exists an open subset U of X such that $A = F \cap U$.*

Proof. In (F, d_F), the open ball of center $x \in F$ and radius $r > 0$ is the set

$$B_F(x,r) = \{y \in F : d(y,x) < r\} = F \cap B(x,r).$$

If $A \subset F$ is open in F, then by definition for any $x \in A$ there exists an $r = r_x > 0$ such that $B_F(x,r_x) = F \cap B(x,r_x) \subset A$. Let $U = \bigcup_{x\in A} B(x,r_x)$; then U is open in X and $F \cap U \subset A$. The reverse inclusion evidently holds, and therefore $A = F \cap U$. Vice versa, assume that $A = F \cap U$ for some open $U \subset X$. Then for any $x \in A$ there exists an $r > 0$ such that $B(x,r) \subset U$; therefore $B_F(x,r) \subset F \cap U = A$, proving that A is open relative to F. □

Examples.
(i) If $A \subset F$ is open (in X), then it is open relative to F.
(ii) $[0,1[$ is open in $[0,\infty[$ and in $[0,2]$, but not in $\mathbb{R}$.
(iii) The set $\{(x,y) \in \mathbb{R}^2 : x \geq 0, y \geq 0, x^2 + y^2 < 1\} = F \cap B(0,1)$ is open in the "first quadrant" $F = \{(x,y) \in \mathbb{R}^2 : x \geq 0, y \geq 0\}$, but not in $\mathbb{R}^2$.

With the help of Proposition 2.6.1 we can easily give examples of disconnected subsets in any metric space X. Indeed, let $F \subset X$ be such that $F = A \cup B$ with $A \neq \emptyset, B \neq \emptyset$ and suppose that A and B can be *separated* by means of two open subsets U and V of X, in the sense that $A \subset U$, $B \subset V$, and $U \cap V = \emptyset$. Then F is disconnected, for A and B are open relative to F; indeed, the stated assumptions imply that $A = F \cap U$, $B = F \cap V$.

Examples.
(i) $F = [0,1] \cup [2,3]$ is a disconnected subset of $\mathbb{R}$.
(ii) $F = [1,2[\cup]2,3]$ is disconnected. More generally, if $J = (a,b) \subset \mathbb{R}$ is any interval and x_0 is any interior point of J, then $F = J\setminus\{x_0\} = (a,x_0[\cup]x_0,b)$ is disconnected; indeed, $(a,x_0[= F \cap]-\infty,x_0[$ is open relative to F, and similarly for $]x_0,b)$.
(iii) $F = \{(x,y) \in \mathbb{R}^2 : xy > 0\}$ is a disconnected subset of $\mathbb{R}^2$, for F is the union of two disjoint non-empty open sets in $\mathbb{R}^2$.

Definition 2.6.2. A metric space X is said to be *connected* if it is not disconnected.

Thus, one way to express the connectedness of X is that X has no subsets which are both open and closed except X itself and the empty set $\emptyset$.

Theorem 2.6.1. *Let $F \subset \mathbb{R}$. Then F is connected if and only if F is an interval.*

Proof. (a) Observe first that F is an interval if and only if for any $x,y \in F$ such that $x < y$ one has $[x,y] \subset F$. Therefore, saying that F is *not* an interval means that there exist points $x_0, y_0 \in F$ and a point z_0 such that $x_0 < z_0 < y_0$ and $z_0 \notin F$. Then put

$$A_{z_0} = F \bigcap]-\infty, z_0[, \quad B_{z_0} = F \bigcap]z_0, +\infty[$$

to obtain two non-empty, disjoint, relatively open subsets of F whose union gives all of F, proving that F is disconnected.

(b) We prove the reverse implication in Theorem 2.6.1 in the case $F = \mathbb{R}$. (The proof in the general case is essentially the same, with the addition of some technicalities.) Assume thus by way of contradiction that $\mathbb{R} = A \cup B$ for some non-empty disjoint open sets $A, B \subset \mathbb{R}$. Pick an $x \in A$ and an $y \in B$ and assume for instance that $x < y$. Put

$$q = \sup\left(A \bigcap [x,y]\right).$$

Then we have

$$x \le q \le y \quad \text{and} \quad q \in \overline{A},$$

the last assertion following by Example 2.2.9. However, A is closed, so $q \in A$. As $y \in B$, it follows first that necessarily $q < y$. Moreover, as A is open, we have $]q - r, q + r[\subset A$ for some $r > 0$, and diminishing r if necessary we can also assume that $q + r < y$. Therefore, $[q, q + r[\subset A \cap [x, y]$, implying that

$$q + r = \sup[q, q + r[\le \sup\left(A \bigcap [x,y]\right) = q,$$

which is absurd. This contradiction proves that $\mathbb{R}$ is connected and ends the proof of Theorem 2.6.1. □

Connectedness and continuity

Theorem 2.6.2. *Let X, Y be metric spaces and let $f : X \to Y$. If f is continuous and X is connected, then $f(X)$ is a connected subset of Y.*

Proof. Suppose by way of contradiction that $f(X)$ is disconnected and let A', B' be relatively open subsets of $f(X)$ such that

$$A' \neq \emptyset,\ B' \neq \emptyset; \quad A' \cap B' = \emptyset; \quad A' \cup B' = f(X).$$

Let $A = f^{-1}(A') = \{x \in X : f(x) \in A'\}$ and similarly let $B = f^{-1}(B')$. Then it is readily verified (see Exercise 2.6.1 below) that

$$A \neq \emptyset,\ B \neq \emptyset; \quad A \cap B = \emptyset; \quad A \cup B = X.$$

Let us now show that A and B are open subsets of X, proving that X is disconnected, contrary to the assumption. Indeed, as A' is open in $f(X)$, there exists an open subset V of Y such that $A' = f(X) \cap V$ (Proposition 2.6.1). Therefore,

$$A = f^{-1}(A') = f^{-1}(f(X) \cap V) = f^{-1}(V)$$

and since f is continuous by assumption, the result follows by the characterization of such property via the inverse images given in Theorem 2.2.3. For the last equality in the formula above, see again Exercise 2.6.1 below. □

Exercise 2.6.1. Let X, Y be any two sets and let f be any map of X into Y. Check that for any subsets H and K of Y one has

$$f^{-1}(H \cup K) = f^{-1}(H) \cup f^{-1}(K) \quad \text{and} \quad f^{-1}(H \cap K) = f^{-1}(H) \cap f^{-1}(K).$$

Deduce from this the following:

(i) Any *partition* of $f(X)$ induces a partition of X via the inverse images through f of the sets of the partition.

(ii) For any set $V \subset X$, one has $f^{-1}(f(X) \cap V) = f^{-1}(V)$.

Exercise 2.6.2. Using Proposition 2.6.1 and (ii) of Exercise 2.6.1, prove that if X, Y are metric spaces, then $f : X \to Y$ is continuous iff it is continuous as a map of X into the metric space $f(X)$.

The two theorems just proved allow to construct further examples of connected spaces and sets.

Definition 2.6.3. A *curve* in a metric space X is a continuous map of an interval $I \subset \mathbb{R}$ into X. If $\gamma : I \to X$ is a curve in X, its image $\gamma(I) \subset X$ is called the *support* of γ. In case $I = [a, b]$, the points $x = \gamma(a), y = \gamma(b)$ are called the *endpoints* of γ and we say that γ *joins* x *and* y *in* X.

In view of Theorems 2.6.1 and 2.6.2, this definition immediately implies the following.

Proposition 2.6.2. *The support of a curve in X is a* connected *subset of X.*

Definition 2.6.4. A metric space X is said to be *path-connected* if any two of its points can be joined by a curve in X; in symbols,

$$\forall x, y \in X \quad \exists [a, b] \subset \mathbb{R} \quad \text{and a continuous} \quad \gamma : [a, b] \to X : x = \gamma(a), y = \gamma(b).$$

Theorem 2.6.3. *A path-connected metric space is connected.*

Proof. Suppose by contradiction that a path-connected space X is the disjoint union of two non-empty open sets A and B. Pick a point $x \in A$ and a point $y \in B$ and join them with a curve γ defined in the interval $[a, b]$, say. Then considering the sets

$$\gamma([a, b]) \cap A, \quad \gamma([a, b]) \cap B$$

we obtain a partition of $\gamma([a, b])$ into two relatively open sets, which contradicts the connectedness of $\gamma([a, b])$ ensured by Proposition 2.6.2. □

Corollary 2.6.1. *The Euclidean space $\mathbb{R}^n$ is connected.*

Indeed, $\mathbb{R}^n$ is path-connected, because any two points $x, y \in \mathbb{R}^n$ can be joined by the *line segment*

$$x + t(y - x), \qquad t \in [0,1].$$

The support of this curve is denoted with the symbol $[x,y]$. Note that in case $n = 1$, this symbol is already used with a different meaning, so we need to verify that they define the same object.

Exercise 2.6.3. Check that if $x, y \in \mathbb{R}$, then

$$\{z = x + t(y - x), t \in [0,1]\} = \{z : x \leq z \leq y\}.$$

Definition 2.6.5. A subset K of $\mathbb{R}^n$ (or, more generally, of a vector space) is said to be *convex* if any two of its points can be joined *in* K with the line segment of endpoints x and y; in symbols,

$$\forall x, y \in K \quad \forall t \in [0,1] \quad x + t(y - x) \in K.$$

Exercise 2.6.4.
(i) Any open (or closed) ball is convex.
(ii) Any vector subspace is convex.
(iii) The intersection of any family of convex sets is convex.

We finally come to the consequences that the connectedness of a domain space X and the continuity of a real-valued function defined on X have on the image set $f(X)$. Loosely speaking, it follows from Theorems 2.6.1 and 2.6.2 that in this case $f(X)$ is one of the intervals (a, b) of endpoints $a = \inf f(x), b = \sup f(x)$ (including the cases $a = -\infty$ or $b = +\infty$). Here are some more precise statements.

Theorem 2.6.4. *Let X be a* connected *metric space and let f be a* continuous *real-valued function defined on X. Then f takes* all *values between* any two *of its values.*

Proof. Let $u, v \in f(X)$ with $u < v$ and let $z \in \mathbb{R}$ be such that $u < z < v$. Since $f(X)$ is an interval, we have $[u, v] \subset f(X)$, so that there is an $x \in X$ such that $f(x) = z$. □

Corollary 2.6.2. *Let $f : X \to \mathbb{R}$ be continuous. If X is connected and f changes sign in X, then f vanishes at some point of X.*

Proof. The assumption is that there exist points x_0 and x_1 in X such that $f(x_0)f(x_1) < 0$. Suppose for instance that $f(x_0) < 0 < f(x_1)$; then it follows by Theorem 2.6.4 that there exists an $x \in X$ such that $f(x) = 0$. □

Corollary 2.6.3. *Let $f : X \to \mathbb{R}$ be continuous. If X is compact and connected, then f takes* all *values between its minimum and maximum values.*

Proof. f attains its minimum and maximum values on X by virtue of the Weierstrass theorem. The conclusion thus follows again from Theorem 2.6.4. □

2.7 Additions and exercises

A1. Further properties of power series

Radius of convergence

Simple methods for the computation of the convergence radius R (Definition 2.1.2) for a given power series – that is, for a given sequence (a_n) of the coefficients in (2.1.10) – are provided by the elementary convergence rules for numerical series. One such criterion says the following.

Lemma 2.7.1. *Let $\sum_0^\infty c_n$ be a numerical series and suppose that the limit $b \equiv \lim_{n\to\infty} |c_n|^{\frac{1}{n}}$ exists (finite or not). Then if $b < 1$, the series converges absolutely, while if $b > 1$, the series does not converge.*

To prove the lemma, we just have to play with the fact that given any $\epsilon > 0$, the inequalities $b - \epsilon < |c_n|^{\frac{1}{n}} < b + \epsilon$ will hold for n large enough. Thus, if $b < 1$ and we pick an $\epsilon > 0$ such that $b + \epsilon < 1$, we will have

$$|c_n| < (b + \epsilon)^n$$

for n larger than some n_0, ensuring the convergence of $\sum_0^\infty |c_n|$ by comparison with (once again!) the geometrical series, while if $b > 1$ and $\epsilon > 0$ is such that $b - \epsilon > 1$, the bound $|c_n| > (b - \epsilon)^n > 1$ shows that c_n cannot tend to zero as $n \to \infty$, and therefore the series cannot converge.

As a consequence of Lemma 2.7.1 we have the following theorem.

Theorem 2.7.1. *Given the series* (2.1.10), *assume that the limit*

$$L = \lim_{n\to\infty} (|a_n|)^{\frac{1}{n}} \tag{2.7.1}$$

exists. Then we have $R = 1/L$, where we put $R = 0$ if $L = +\infty$ and $R = +\infty$ if $L = 0$.

Proof. Simply apply Lemma 2.7.1 to (2.1.10), noting that

$$\lim_{n\to\infty} |a_n x^n|^{\frac{1}{n}} = L|x|.$$

Indeed, consider first the case $0 < L < +\infty$. Then if $|x| < 1/L$, we have $L|x| < 1$, so by Lemma 1.1.1, the series $\sum_0^\infty a_n x^n$ converges, while if $|x| > 1/L$, we have $L|x| > 1$ and therefore the series will not converge. The extreme cases $L = 0$ and $L = +\infty$ can be dealt with in a similar way. □

Exercise 2.7.1. Compute the radii of convergence of the following series:

$$\sum_{n=0}^{+\infty} \frac{x^n}{n^a} \; (a \in \mathbb{R}), \quad \sum_1^\infty \frac{x^n}{a^n + b^n} \; (a > 0, b > 0), \quad \sum_{n=1}^\infty \frac{2^n - 1}{n^2 + \sqrt{n} + 10} x^n.$$

Regularity properties of the function sum of a power series

Given the series (2.1.10) with a radius of convergence $R > 0$, we can consider the properties of the function defined putting

$$f(x) = \sum_0^\infty a_n x^n \quad (-R < x < R). \tag{2.7.2}$$

From Theorem 2.1.4 and Definition 2.1.2 it follows that the series in (2.7.2) converges uniformly on each interval $[-r, r]$ with $0 < r < R$. This implies in particular that f is **continuous** in the whole open interval $]-R, R[$ and that we can integrate term-by-term (Corollary 2.1.2) a power series in each interval $[a, b] \subset]-R, R[$; in particular, we have

$$\int_0^x f(t)\,dt = a_0 x + a_1 \frac{x^2}{2} + \cdots a_n \frac{x^{n+1}}{n+1} + \cdots \tag{2.7.3}$$

for every x with $|x| < R$. For instance, the equality $1/(1+x) = \sum_0^\infty (-1)^n x^n$ implies that

$$\log(1+x) = \sum_0^\infty (-1)^n \frac{x^{n+1}}{n+1} = x - \frac{x^2}{2} + \frac{x^3}{3} + \cdots \quad (|x| < 1).$$

We are now going to see – using the "series derivation" Corollary 2.1.3 – that the sum of a power series has much stronger regularity properties than mere continuity.

Theorem 2.7.2. *Let f be as in* (2.7.2). *Then $f \in C^\infty(]-R, R[)$ and, for every $k \in \mathbb{N}$, we have*

$$(D^k f)(x) = \sum_{n=k}^\infty n(n-1)\dots(n-k+1) a_n x^{n-k}. \tag{2.7.4}$$

Proof. Fix $k \in \mathbb{N}$. Note that the expression appearing under the sum in (2.7.4) is nothing but the kth derivative of the nth term $f_n(x) \equiv a_n x^n$ $(n \geq k)$ of the power series; thus, our aim is to prove that for $|x| < R$ we have

$$(D^k f)(x) = \sum_{n=k}^\infty (D^k f_n)(x) = \sum_{n=0}^\infty (D^k f_n)(x)$$

(note that $D^k(a_n x^n) = 0$ if $n < k$). However, this follows from a repeated use of Corollary 2.1.3 once we note that the series on the right-hand side of (2.7.4) has the same radius of convergence R as the original series in (2.7.2). To see this, multiply each term of the series in (2.7.4) by the fixed term x^k – which will not change its radius of convergence – to obtain the conveniently modified series

$$\sum_{n=k}^\infty n(n-1)\dots(n-k+1) a_n x^n \tag{2.7.5}$$

and now use Theorem 2.7.1 to compute its radius of convergence; recalling that

$$\lim_{n\to\infty} n^{\frac{1}{n}} = \cdots = \lim_{n\to\infty} (n-k+1)^{\frac{1}{n}} = 1$$

we obtain the equality

$$\lim_{n\to\infty} (n(n-1)\dots(n-k+1)|a_n|)^{\frac{1}{n}} = \lim_{n\to\infty} |a_n|^{\frac{1}{n}} = R,$$

proving our claim and thus proving Theorem 2.7.2. □

Example 2.7.1. Theorem 2.7.2 permits to differentiate term-by-term a power series as many times as we like within its interval of convergence. For instance, from the equality $1/(1-x) = \sum_{k=0}^{\infty} x^k$ we derive

$$\frac{1}{(1-x)^2} = \sum_{k=1}^{\infty} kx^{k-1} = \sum_{k=0}^{\infty} (k+1)x^k = 1 + 2x + 3x^2 + \cdots \quad (|x| < 1),$$

and iterating this procedure, we can expand $1/(1-x)^3$, and so on.

Taylor expansion

Putting $x = 0$ in (2.7.4) – so that only the first term does not vanish – we see that

$$a_k = \frac{D^k f(0)}{k!} = \frac{f^{(k)}(0)}{k!} \quad \forall k \in \mathbb{N}.$$

Given a function f of class C^∞ in a neighborhood $I =]-a, a[$ of $x = 0$, the series

$$\sum_{k=0}^{\infty} \frac{f^{(k)}(0)}{k!} x^k$$

is called the **Taylor series of f** centered at 0, and a main question is to see if it converges to f in I (or a smaller neighborhood of $x = 0$). Simple examples show that this is not always true, and we conclude this quick account reporting a simple criterion ensuring that the **power series expansion** of a C^∞ function takes place.

Theorem 2.7.3. *Let $f \in C^\infty(]-r, r[)$. Suppose that f has **uniformly bounded derivatives** in $(]-r, r[)$ in the sense that there is an $M > 0$ such that*

$$|f^{(n)}(x)| \le M \quad \forall x \in]-r, r[, \quad \forall n \in \mathbb{N}. \tag{2.7.6}$$

Then f can be expanded in power series in $]-r, r[$; that is to say, we have

$$f(x) = \sum_{n=0}^{\infty} \frac{f^{(n)}(0)}{n!} x^n \quad \forall x \in]-r, r[. \tag{2.7.7}$$

Proof. It is a matter of using **Taylor's formula** for f, which tells us that for every $n \in \mathbb{N}$ we have

$$f(x) = \sum_{k=0}^{n} \frac{f^{(k)}(0)}{k!} x^k + R_n(x) \quad \forall x \in]-r, r[, \tag{2.7.8}$$

where the **remainder term** $R_n(x)$ is $o(|x|^n)$ as $x \to 0$, meaning that $R_n(x)/|x|^n$ tends to 0 as $n \to \infty$. It is convenient to use the **Lagrange expression** for the remainder, which tells us that for some $z \in]-r, r[$,

$$R_n(x) = \frac{f^{(n+1)}(z)}{(n+1)!} x^{n+1}.$$

Now our assumption (2.7.6) implies that

$$|R_n(x)| \le M \frac{r^{n+1}}{(n+1)!} \to 0 \quad \text{as } n \to \infty$$

(uniformly for $x \in]-r, r[$), so that the conclusion (2.7.7) follows on letting $n \to \infty$ in Taylor's formula (2.7.8). □

Example 2.7.2. The condition (2.7.6) applies of course to the functions $\sin x$ and $\cos x$ with $r = \infty$ and thus justifies their well-known Taylor expansions on $\mathbb{R}$. But it also applies to the function e^x, which does not have bounded derivatives on all of $\mathbb{R}$, but does so on each arbitrarily fixed interval $]-r, r[$. This allows for the expansion

$$e^x = \sum_{n=0}^{\infty} \frac{x^n}{n!} = 1 + x + \frac{x^2}{2!} + \cdots, \tag{2.7.9}$$

which is valid for every $x \in \mathbb{R}$ and is used as starting point to define and deal with the exponential matrix e^A (Chapter 1, Section 1.4).

A2. Equivalent norms in a vector space

Definition 2.7.1. Given a vector space E, two norms $\|\cdot\|_1$ and $\|\cdot\|_2$ in E are said to be **equivalent** if there exist constants $c > 0, d > 0$ such that

$$c\|x\|_1 \le \|x\|_2 \le d\|x\|_1 \tag{2.7.10}$$

for every $x \in E$.

The same definition could clearly be given, more generally, for two distances on any set.

Exercise 2.7.2. Prove that if two norms on a vector space E are equivalent, then the open sets of $E_1 \equiv (E, \|\cdot\|_1)$ and $E_2 \equiv (E, \|\cdot\|_2)$ are the same and thus define the same topology on E.

Theorem 2.7.4. *If E is a* finite-dimensional *vector space, then any two norms on E are equivalent.*

Theorem 2.7.4 is a consequence of the following structural fact concerning finite-dimensional normed vector spaces. If $\dim E = n$ and $\{v_1, \dots, v_n\}$ is a basis of E, we know by the definition of basis that the linear map

$$H : \mathbb{R}^n \to E : x = (x_1, \dots, x_n) \to x_1 v_1 + \cdots x_n v_n$$

is bijective and hence an isomorphism of $\mathbb{R}^n$ onto E in the purely algebraic sense; in fact, we say briefly that every n-dimensional vector space is isomorphic to $\mathbb{R}^n$. If we put a norm $\|\cdot\|$ on E, what can we say about the **continuity** of H and H^{-1}? The continuity of H is straightforward to check, for we have (denoting here by $\|\cdot\|_2$ the Euclidean norm in $\mathbb{R}^n$)

$$\|H(x)\| = \left\| \sum_{i=1}^{n} x_i v_i \right\| \le \sum_{i=1}^{n} |x_i| \|v_i\| \le \|x\|_2 \sum_{i=1}^{n} \|v_i\| \equiv K \|x\|_2 \quad (x \in \mathbb{R}^n). \tag{2.7.11}$$

The proof of the continuity of H^{-1} is more delicate and can be done for instance by induction on the dimension n of E (see Dieudonné [7]). A bijective map between metric (or more generally, topological) spaces that is continuous together with its inverse is called a *homeomorphism*. We can therefore state the following.

Theorem 2.7.5. *Every n-dimensional normed vector space is* ***linearly homeomorphic*** *to $\mathbb{R}^n$.*

Some authors (for instance, Taylor [8]) use the term *topological isomorphism* rather than linear homeomorphism.

Exercise 2.7.3. Prove (without making reference to the theorems stated above) that in $\mathbb{R}^n$ any two norms are equivalent.

Hint: It is clearly enough to prove that any norm $\|\cdot\|$ in $\mathbb{R}^n$ is equivalent to the Euclidean norm $\|\cdot\|_2$. To see this, first check (as in (2.7.11)) that $\|x\| \le C\|x\|_2$ for some $C > 0$ and all $x \in \mathbb{R}^n$. Deduce from this that

$$\big| \|x\| - \|y\| \big| \le C\|x - y\|_2 \quad (x, y \in \mathbb{R}^n)$$

and finally apply the Weierstrass theorem (Theorem 2.5.7) to conclude that $\|x\| \ge c$ for some $c > 0$ and all $x \in S \equiv \{x \in \mathbb{R}^n : \|x\|_2 = 1\}$, whence it follows that $\|x\| \ge c\|x\|_2$ for every x, giving the result.

A3. Further examples of bounded linear Operators

1. (Linear operators on finite-dimensional spaces) Let T be a linear map of a finite-dimensional normed vector space E into a normed vector space F. Then T is bounded (in the sense of Example 2.2.7). Indeed, assuming $\dim E = n$, fix a basis $v_1, \dots, v_n$ of E and represent every $x \in E$ in this basis writing $x = x_1v_1 + \cdots x_nv_n$; also put

$$\|x\|_1 \equiv \sum_{i=1}^{n} |x_i|.$$

Then we have

$$\begin{aligned}\|T(x) &= \|T(x_1v_1 + \cdots + x_nv_n\| = \|x_1T(v_1) + \cdots + x_nT(v_n)\| \\ &\le |x_1|\|T(v_1)\| + \cdots |x_n|\|T(v_n)\| \le K\sum_{i=1}^{n}|x_i| = K\|x\|_1 \le K'\|x\|,\end{aligned}$$

where $K = \sum_{i=1}^{n} \|T(v_i)\|$ and we have used the fact that in E all norms are equivalent by Theorem 2.7.4.

2. (Linear integral operators) Let $E = C([a,b])$ equipped with the sup norm $\|x\| = \sup_{a\le t\le b}|x(t)|$ for $x \in E$ and let T be the linear operator of E into itself defined putting

$$T(x)(s) = \int_{c}^{d} k(s,t)x(t)\,dt \quad (a \le s \le b) \tag{2.7.12}$$

for every $x \in E$, where k is a continuous real-valued function defined in the rectangle $R = [a,b]\times[c,d]$, called the **kernel** of the integral operator T. To check that the function $T(x)$ defined on $[a,b]$ through formula (2.7.12) is continuous (i. e., belongs to E), consider that the continuity of k and the compactness of $R = [a,b] \times [c,d]$ imply the uniform continuity of k on R by the Heine–Cantor theorem (Theorem 2.5.8). In particular, this ensures that given any $\epsilon > 0$, there is a $\delta > 0$ so that

$$\forall s,s' \in [a,b], \quad |s-s'| < \delta \Rightarrow |k(s,t) - k(s',t)| < \epsilon \quad \forall t \in [c,d]. \tag{2.7.13}$$

Using (2.7.13) in (2.7.12), the continuity of $T(x)$ follows immediately.

To check that the linear operator $T : E \to E$ is bounded, write

$$|T(x)(s)| \le \int_{c}^{d} |k(s,t)||x(t)|\,dt \le M\int_{c}^{d}|x(t)|\,dt \le M(d-c)\|x\|, \tag{2.7.14}$$

where $M = \sup_{(s,t)\in R}|k(s,t)|$; and as (2.7.14) holds for every $s \in [a,b]$, we conclude that

$$\|T(x)\| \le M(d-c)\|x\| \quad (x \in E).$$

3. (Compact linear operators) Let E, F be normed linear spaces and let $T : E \to F$ be a bounded linear operator. T is said to be **compact** if for every bounded sequence $(x_n) \subset E$, there is a subsequence (x_{n_k}) such that $T(x_{n_k})$ converges in F. Some remarks about this definition follow:

(i) Using the definitions and results given in Section 2.5, it is not hard to check that the property stated above is equivalent to the fact that for every bounded subset B of E, the image set $T(B)$ is (not only bounded, but also) **relatively compact**, meaning that the closure $\overline{T(B)}$ is compact. Since every compact set is bounded, we can then first note that there is no need to specify in advance that T is a *bounded* linear operator; in other words, compactness of a linear operator implies *per se* its boundedness.

(ii) Every bounded linear operator of a normed space E into a finite-dimensional normed space F is compact. Indeed, assume first that $F = \mathbb{R}^n$ and observe that for any bounded subset B of E, the subset $T(B) \subset \mathbb{R}^n$ will be bounded by the assumption on T and therefore also relatively compact by Theorem 2.5.4. To deal with the general case, use Theorem 2.7.5.

Example 2.7.3. The linear integral operator T acting in $E = C([a,b])$ defined in (2.7.12) is compact. This fact rests on a fundamental result in function theory, the **Ascoli–Arzelà theorem**, which characterizes the relatively compact subsets B of $C([a,b])$ asking that they should be bounded and **equicontinuous**, meaning that given $\epsilon > 0$ there exists a $\delta > 0$ such that for every $s, t \in [a,b]$ with $|s-t| < \delta$ and **every** $x \in B$ we have

$$\big|x(s) - x(t)\big| < \epsilon.$$

References for the statement and proof of Ascoli–Arzelà's theorem are for instance Dieudonné [7] and Kolmogorov–Fomin [9]. To show that the set $T(B)$ is equicontinuous if $B \subset E$ is bounded, we use again the uniform continuity of k in $[a,b] \times [c,d]$ to ensure that given $\epsilon > 0$, there is a $\delta > 0$ so that (2.7.13) holds. Then the definition (2.7.12) of T shows that if $|s - s'| < \delta$,

$$\big|T(x)(s) - T(x)(s')\big| \le \int_c^d \big|k(s,t) - k(s',t)\big\|x(t)\big|\,dt \le \epsilon \int_c^d \big|x(t)\big|dt \le \epsilon(d-c)\|x\|$$

and thus proves that the functions of the family $T(B)$ are equicontinuous, since $\|x\| \le C$ for some $C > 0$ and for all $x \in B$.

4. (Orthogonal projection onto a closed subspace) As we shall see in the next chapter (Corollary 3.4.2), given a closed vector subspace M of a Hilbert space H, it is possible to decompose every vector x of H in the form

$$x = y + z, \quad y \in M,\ z \perp M, \tag{2.7.15}$$

where $z \perp M$ means that the scalar product $\langle z, v\rangle = 0$ for every $v \in M$; moreover, the decomposition in (2.7.15) is unique. The vector $y \in M$ corresponding to x is the **orthogonal projection** of x onto M and denoted for instance Px; this defines a map P of H into H. Using the uniqueness of the decomposition as in (2.7.15) and the Pythagoras identity (3.1.4), it is easily seen that P is a bounded linear operator in H.

Exercises

There are many exercises spread out over the various sections of this chapter. Purposely, no solutions are given here, as the reader of the present chapter is invited to solve them entirely by him/herself. A few additional exercises are given below.

Further exercises

Exercise 2.7.4. Given the series

$$\sum_{1}^{\infty} \arctan x^n \quad (x \in \mathbb{R}),$$

first find its (pointwise) convergence set A and then show that the sum function is of class C^1 in A.

Exercise 2.7.5.

(i) Let $f : \mathbb{R}^n \to \mathbb{R}$ be continuous. Prove that if $f(x_0) > 0$ for some x_0, then $f(x) > 0$ for every x in some neighborhood of x_0. Also prove that if $f(x) \neq 0$ for every $x \in \mathbb{R}^n$, then either $f(x) > 0$ for every $x \in \mathbb{R}^n$ or $f(x) < 0$ for every $x \in \mathbb{R}^n$.

(ii) Show that the set $A = \{\frac{1}{n} : n \in \mathbb{N}\}$ is not closed in $\mathbb{R}$, while the rectangle $R = [0,1] \times [0,1]$ is closed in $\mathbb{R}^2$.

(iii) If R is as in (ii), determine the image $f(R)$ of R through the function $f : \mathbb{R}^2 \to \mathbb{R}$ defined putting $f(x,y) = \sqrt{x^2 + y^2}e^{xy}$. For the same f, determine the preimage $f^{-1}(V)$ with $V =]0, +\infty[$ and with $V = [-1, 0]$.

(iv) Is the set

$$A = \{(x,y) \in \mathbb{R}^2 : e^{xy} > 1, y \le -1\}$$

open? Is it compact? Is it connected? Explain.

Exercise 2.7.6. Let $(E, \|.\|)$ be a normed vector space. Given $x_0 \in E$ and $r > 0$, let $x \in E$ be such that $\|x - x_0\| = r$. Putting

$$x_n = x_0 + t_n(x - x_0), \quad \text{where } 0 < t_n < 1 \quad \forall n \in \mathbb{N},$$

check that $(x_n) \subset B(x_0, r)$ and that $x_n \to x$ if $t_n \to 1$. Deduce from this that $x \in \overline{B(x_0, r)}$ and conclude that

$$\overline{B(x_0, r)} = B'(x_0, r) \equiv \{x \in \mathbb{E} : \|x - x_0\| \le r\}.$$

Exercise 2.7.7. Let $E = C([a,b])$ with the sup norm $\|x\| = \sup_{a \le t \le b} |x(t)|$ for $x \in E$. Check that if $f : A \equiv [a,b] \times \mathbb{R} \to \mathbb{R}$ is continuous in A and Lipschitzian with respect to the second variable in A (Definition 1.1.2), then the Nemytskii operator N_f induced by f and defined putting

$$N_f(u)(t) = f(t, u(t)) \quad \text{for } u \in E \text{ and } t \in [a, b]$$

(Remark 1.1.1) is a Lipschitz map from E to E. Does this still hold if in E the uniform norm $\|\cdot\|$ is replaced with the integral norm $\|\cdot\|_1$ defined in (2.2.3)?

Exercise 2.7.8. Use Theorem 2.7.5 to prove that every finite-dimensional normed space is a Banach space.

3 Fourier series and Hilbert spaces

Introduction

A good part of this chapter – beginning with Section 3.1 – is devoted to showing the advantages of having an inner product in a vector space: this permits to approach in an abstract setting concepts like that of *orthogonality* and consequently that of *projection onto a subspace*, which closely remind us of the concept of *perpendicularity* and that of *projection onto a line* (or a plane), which we learn from the elementary geometry of the three-dimensional space and then review in linear algebra.

The concept of orthogonal projection onto a subspace (Section 3.1) establishes a strict and efficient connection between the *geometric* notion of orthogonality and the *metric* notion of nearest point in a subspace to a given point. More explicitly, given a subspace M of an inner product space E and given $x \in E$, the characterization of the orthogonal projection of x onto M as the unique $y \in M$ such that

$$x - y \perp M \quad \text{or equivalently} \quad \|x - y\| \leq \|x - z\| \quad \forall z \in M$$

proves to be extremely fruitful and is used at several points throughout the whole chapter, beginning with Section 3.2, where it is used to characterize the finite sums of the Fourier series of a periodic f as the trigonometric polynomials of best approximation to f in the quadratic mean, which is a crucial point for the convergence of the Fourier series of f to f itself.

Section 3.3 is devoted to the study of orthonormal systems and their special properties, leading to famous results such as Bessel's inequality or Parseval's identity, which are frequently used in applications and especially in approximation theory. Due evidence is given to the property of *totality*, or *completeness*, of an orthonormal system (e_n) in an inner product space E and to the consequent possibility of expanding every vector $x \in E$ in the form

$$x = \sum_{n=1}^{\infty} \langle x, e_n \rangle e_n. \quad (*)$$

Finally, in Section 3.4, we let the metric completeness come in as a final ingredient to present the Hilbert spaces. In order to illustrate their importance and use, we first give a rapid description of the space L^2 of square-integrable functions (and more generally of the L^p spaces with $1 \leq p < \infty$) and, before that, of the more intuitive sequence space l^2. As a next step, besides discussing basic results such as the Fischer–Riesz theorem, we focus especially on the existence and uniqueness of the nearest point in a closed convex set to a given point, which greatly extends that of the orthogonal projection onto a subspace. Besides showing once more the power of the inner product, this result can be taken as a starting point for the study of minimization problems and of the role of *convexity* in this context.

https://doi.org/10.1515/9783111302522-003

In the additional Section 3.5, aside from discussing as usual a few exercises, we take some space to present the spectral theory of Sturm–Liouville operators, with the aim of indicating their connection with the study of Fourier or more general expansions like (*) as *eigenfunction* expansions, in view of their importance for the method of separation of variables, which will be concretely used in the problems of PDEs to be studied in the final Chapter 4.

3.1 Inner product spaces. Orthogonal projection onto a subspace

Definition 3.1.1. Let E be a real vector space. An *inner product* (or *scalar product*) in E is a mapping s of $E \times E$ into $\mathbb{R}$ such that the following properties hold for every $x, y, z \in E$ and every $\alpha \in \mathbb{R}$:
(i) $s(x+y,z) = s(x,z) + s(y,z)$,
(ii) $s(\alpha x, y) = \alpha s(x,y)$,
(iii) $s(x,y) = s(y,x)$,
(iv) $s(x,x) \geq 0$ and $s(x,x) = 0$ if and only if $x = 0$.

In words, we say that s is a *positive definite, symmetric, bilinear form on* $E \times E$. A vector space equipped with an inner product is called an *inner product space.*

Note that the bilinearity of s is a consequence of its linearity in the first argument and its symmetry. That is to say, using properties (i), (ii), and (iii) we also have

$$s(x, \alpha y + \beta z) = \alpha s(x,y) + \beta s(x,z)$$

for every $x, y, z \in E$ and every $\alpha, \beta \in \mathbb{R}$.

Given two vectors $x, y \in E$, the real number $s(x,y)$ is called the *inner product between* x *and* y.

Notation. We shall write $\langle x, y \rangle$ instead of $s(x,y)$.

Example 3.1.1. Let $E = \mathbb{R}^n$. For $x = (x_1, \dots, x_n), y = (y_1, \dots, y_n) \in \mathbb{R}^n$, put

$$\langle x, y \rangle = \sum_{i=1}^{n} x_i y_i. \tag{3.1.1}$$

Example 3.1.2. Let $E = C([a,b])$. For $f, g \in E$, put

$$\langle f, g \rangle = \int_a^b f(x) g(x)\, dx. \tag{3.1.2}$$

Exercise 3.1.1. Check that (3.1.2) defines an inner product in $C([a,b])$. To verify property (iv) in Definition 3.1.1, the following proposition is needed.

Proposition 3.1.1. *Let $g \in C([a,b])$ with $g \geq 0$ in $[a,b]$. If $\int_a^b g(x)dx = 0$, then $g(x) = 0$ for every $x \in [a,b]$.*

The Cauchy–Schwarz inequality

Proposition 3.1.2. *Let E be a vector space with inner product $\langle .,. \rangle$. Then*

$$|\langle x,y \rangle| \leq \sqrt{\langle x,x \rangle}\sqrt{\langle y,y \rangle} \tag{3.1.3}$$

for every $x, y \in E$.

Proof. Given any $x, y \in E$, using properties (i)–(iv) of Definition 3.1.1 we have for any $t \in \mathbb{R}$

$$\begin{aligned}\langle x - ty, x - ty \rangle &= \langle x, x - ty \rangle + \langle -ty, x - ty \rangle = \langle x,x \rangle - t\langle x,y \rangle - t\langle y,x \rangle + t^2\langle y,y \rangle \\ &= \langle x,x \rangle - 2t\langle x,y \rangle + t^2\langle y,y \rangle \\ &\equiv A - 2tB + t^2C \geq 0.\end{aligned}$$

It then follows that

$$B^2 \leq AC$$

or equivalently that $|B| \leq \sqrt{A}\sqrt{C}$, which is (3.1.3). □

Proposition 3.1.3. *Let E be a vector space with inner product $\langle .,. \rangle$. Then putting*

$$\|x\| = \sqrt{\langle x,x \rangle} \tag{3.1.4}$$

we have a norm *on E, which is said to be* ***induced by the inner product****.*

Proof. First note that the definition (3.1.4) makes sense by virtue of property (iv) of the inner product. Likewise, using also properties (ii) and (iii), it follows at once that
- $\|x\| \geq 0\ \forall x \in E$ and $\|x\| = 0 \Leftrightarrow x = 0$;
- $\|\alpha x\| = |\alpha|\|x\|\ \forall x \in E,\ \forall \alpha \in \mathbb{R}$.

As to the triangle inequality, given any $x, y \in E$, we have by the definition (3.1.4)

$$\|x + y\|^2 = \langle x + y, x + y \rangle = \langle x,x \rangle + 2\langle x,y \rangle + \langle y,y \rangle,$$

whence, using the Cauchy–Schwarz inequality (3.1.3), it follows that

$$\|x + y\|^2 \leq \|x\|^2 + 2\|x\|\|y\| + \|y\|^2 = (\|x\| + \|y\|)^2,$$

so that finally

$$\|x + y\| \leq \|x\| + \|y\|.$$

□

Examples. The norm induced on $E = \mathbb{R}^n$ by the scalar product (3.1.1) is the *Euclidean norm*

$$\|x\| = \sqrt{\sum_{i=1}^{n} x_i^2} \equiv \|x\|_2 \tag{3.1.5}$$

(which has to be distinguished by other possible norms on $\mathbb{R}^n$, such as for instance $\|x\|_1 \equiv \sum_{i=1}^{n} |x_i|$).

The norm induced on $E = C([a,b])$ by the scalar product (3.1.2) is

$$\|f\| = \sqrt{\int_a^b f^2(x)\,dx} = \left(\int_a^b f^2(x)\,dx\right)^{1/2} \equiv \|f\|_2 \tag{3.1.6}$$

and has to be distinguished by other norms on $C([a,b])$, such as for instance the norm $\|f\| = \sup_{x\in[a,b]} |f(x)|$ of uniform convergence, mostly considered in Chapter 2 and often denoted $\|f\|_\infty$.

Remark 3.1.1. It follows from Proposition 3.1.3 that any inner product space E becomes in a canonical way a normed space, and thus in turn – as any normed space, see Example 2.2.2 – it has to be considered a **metric space** with the **distance induced by the norm**, defined putting

$$d(x,y) = \|x-y\| \quad (x,y \in E).$$

Therefore, every concept known for metric spaces (neighborhoods, open and closed sets, convergent and Cauchy sequences, compactness, continuity – in a word, the **topology** of metric spaces) applies in particular to inner product spaces using the norm (3.1.4).

Remark 3.1.2. It follows in particular that when $C([a,b])$ is equipped with the inner product (3.1.2), the distance between two functions $f,g \in C([a,b])$ is the **distance in the quadratic mean**

$$d_2(f,g) = \|f-g\|_2 = \left(\int_a^b |f(x)-g(x)|^2\,dx\right)^{1/2}.$$

Therefore, the convergence of a sequence $(f_n) \subset C([a,b])$ to an $f \in C([a,b])$ in this normed space means that

$$d_2(f_n,f) = \|f_n - f\|_2 = \left(\int_a^b |f_n(x)-f(x)|^2\,dx\right)^{1/2} \to 0 \quad (n\to\infty)$$

and is expressed by saying that (f_n) **converges to** f **in the quadratic mean**, or *in the sense of least squares* (see, e. g., Weinberger [2]).

Orthogonality

Definition 3.1.2. Let E be a vector space with inner product $\langle .,. \rangle$. Two vectors $x, y \in E$ are said to be *orthogonal* if

$$\langle x, y \rangle = 0. \tag{3.1.7}$$

We also say that x *is orthogonal to* y (which is the same as saying that y is orthogonal to x) and write

$$x \perp y.$$

It follows by the definition of inner product (Definition 3.1.1) that:

- every vector $x \in E$ is orthogonal to the vector 0;
- the only vector orthogonal to itself is the vector $x = 0$, whence it follows that the only vector orthogonal to *all* vectors of E is $x = 0$.

Proposition 3.1.4. *Let E be a vector space with inner product $\langle .,. \rangle$ and let x, y be any two vectors of E. Then*

$$\|x + y\|^2 + \|x - y\|^2 = 2\|x\|^2 + 2\|y\|^2. \tag{3.1.8}$$

Moreover, if x, y are orthogonal, then

$$\|x + y\|^2 = \|x\|^2 + \|y\|^2. \tag{3.1.9}$$

Proof. This is left as an exercise.

Equalities (3.1.8) and (3.1.9) are called respectively the **parallelogram law** and the **Pythagoras identity**.

The concept of orthogonality can be extended to any (finite or infinite) sequence (v_i) of vectors of E, meaning that

$$\langle v_i, v_j \rangle = 0 \quad \forall i \neq j.$$ □

Proposition 3.1.5. *If $(v_i)_{1 \leq i \leq n}$ are orthogonal vectors of an inner product space E with $v_i \neq 0$ $(1 \leq i \leq n)$, then they are linearly independent.*

Proof. This is left as an exercise. □

Definition 3.1.3. A (finite or infinite) sequence $(e_n)_{n=1,2,\dots}$ of vectors of an inner product space E is said to be **orthonormal** if

$$\begin{cases} \langle e_i, e_j \rangle = 0 & (i \neq j) \\ \langle e_i, e_i \rangle = 1. \end{cases} \tag{3.1.10}$$

We also say that the vectors $(e_n)_{n=1,2,\dots}$ are orthonormal.

Thus, orthonormal vectors are norm-one eigenvectors that are orthogonal to each other. It is also clear that any sequence (v_i) of orthogonal vectors with $v_i \neq 0$ for every i can be turned into a sequence of orthonormal vectors just **normalizing** them, that is, considering the vectors $u_i = v_i/\|v_i\|$.

Orthogonal projection onto a finite-dimensional subspace

Theorem 3.1.1. *Let $e_1, \ldots, e_n$ be orthonormal vectors of an inner product space E and let $M = [e_1, \ldots, e_n]$ denote the vector subspace spanned by these vectors. Then for every $x \in E$, there is a unique $y \in M$ such that*

$$x - y \perp M \tag{3.1.11}$$

and we have the explicit expression

$$y = \sum_{i=1}^{n} \langle x, e_i \rangle e_i. \tag{3.1.12}$$

Moreover, y satisfies the inequality

$$\|x - y\| \leq \|x - z\| \quad \forall z \in M, \tag{3.1.13}$$

where strict inequality holds for $z \neq y$. Finally, we have

$$\|x - y\|^2 = \|x\|^2 - \sum_{i=1}^{n} \langle x, e_i \rangle^2. \tag{3.1.14}$$

*The vector y is called the **orthogonal projection of x onto the subspace** M.*

Proof. (i) The notation $v \perp M$ means (of course) that $v \perp z$ for every $z \in M$; it can be used for any subset M of E, and not only for vector subspaces. Given $x \in E$, we prove at once the existence and uniqueness of y with the property (3.1.11), as well as its explicit expression (3.1.12), as follows. As $y \in M$, we need to have $y = \sum_{i=1}^{n} c_i e_i$ for some $c_i \in \mathbb{R}$ $(1 \leq i \leq n)$ to be determined; moreover, the condition (3.1.11) will be satisfied iff $x - y$ is orthogonal to each of the spanning vectors e_i of M. Therefore, we must have

$$\left\langle x - \sum_{i=1}^{n} c_i e_i, e_j \right\rangle = 0 \quad \forall j = 1, \ldots, n,$$

whence – using the orthonormality relations (3.1.10) – it follows that

$$\langle x, e_j \rangle = c_j \quad \forall j = 1, \ldots, n,$$

proving the first statement of Theorem 3.1.1.

(ii) To prove the minimal property (3.1.13), take any $z \in M$ and write

$$\|x-z\|^2 = \|x-y+y-z\|^2 = \|x-y\|^2 + \|y-z\|^2 \geq \|x-y\|^2,$$

the second equality being justified by (3.1.9), for $x - y \perp M$ while $y - z \in M$.

(iii) Finally, to prove (3.1.14), write – using once again (3.1.11) –

$$\|x-y\|^2 = \langle x-y, x-y\rangle = \langle x, x-y\rangle = \langle x,x\rangle - \langle x,y\rangle.$$

Moreover, by (3.1.12) we have

$$\langle x,y\rangle = \left\langle x, \sum_{i=1}^{n} \langle x, e_i\rangle e_i \right\rangle = \sum_{i=1}^{n} \langle x, e_i\rangle^2$$

so that the result follows. □

Additional remarks

Remark 3.1.3. Given $x \in E$, the **uniqueness** of a $y \in M$ with the orthogonality property (3.1.11) holds independently of the representation of M by means of an orthonormal basis and holds in fact if M is **any** vector subspace of E. Let us state this formally.

Proposition 3.1.6. *Let $x \in E$ and let M be a vector subspace of E. Then there exists **at most one** $y \in M$ such that*

$$x - y \perp M.$$

Proof. Let $x \in E$ and suppose that $y_1, y_2 \in M$ satisfy the property above. Then, by linearity of the inner product,

$$x - y_1 - (x - y_2) = y_2 - y_1 \perp M.$$

However, $y_2 - y_1 \in M$ because M is a subspace; thus, $y_2 - y_1 \perp y_2 - y_1$, whence $y_2 = y_1$. □

Remark 3.1.4. The "optimality" property (3.1.13) says that given $x \in E$ (and a subspace M of E), its orthogonal projection y on M has a distance from x that is minimal with respect to any other point z of M; briefly, y is the **nearest point to** x **in** M. It is important to know that this property *characterizes* the orthogonal projection of x on M, that is to say, if $y \in M$ satisfies (3.1.13), then necessarily

$$x - y \perp M.$$

Thus, we conclude that the orthogonality property (3.1.11) and the nearest point property (3.1.13) are **equivalent** to each other. Again we state this formally.

Proposition 3.1.7. *Let $x \in E$ and let M be a vector subspace of E. If $y \in M$ has the nearest point property* (3.1.13), *that is, it is such that*

$$\|x - y\| \leq \|x - z\| \quad \forall z \in M, \tag{3.1.15}$$

then necessarily $x - y \perp M$.

Proof. Let $v \in M$. We need to prove that $\langle x - y, v \rangle = 0$. For this purpose, consider that for every $t \in \mathbb{R}$ the vector $y + tv \in M$; therefore, by (3.1.15) we have

$$\begin{aligned}\|x - y\|^2 &\leq \|x - (y + tv)\|^2 \\ &= \|(x - y) - tv\|^2 = \|x - y\|^2 - 2t\langle x - y, v \rangle + t^2\|v\|^2 \equiv h(t).\end{aligned}$$

The inequality above shows that the function h has an absolute minimum at $t = 0$; therefore,

$$h'(0) = \left(2t\|v\|^2 - 2\langle x - y, v \rangle\right)_{t=0} = -2\langle x - y, v \rangle = 0.$$ □

Some questions

- Given $x \in E$ and **any** vector subspace M of E, can one prove the **existence** of the orthogonal projection of x onto M – that is, of an $y \in M$ enjoying the (equivalent) properties (3.1.11) and (3.1.13)?
- Given $x \in E$ and a subset $K \subset E$, can we prove the existence of a nearest point in K to x for more general sets K than just vector subspaces?

As we shall see in Section 3.4 of the present chapter, an affirmative answer to the first question can be given if we require the **completeness** of the inner product space E – that is, if we require that E be what is called a **Hilbert space** – and the **closedness** of the subspace M.

The second question is, in a sense, just a more general version of the first, and indeed they will be answered together in the context of Hilbert spaces, giving full evidence of the importance of the **convexity** (and closedness) of the set K. For the time being, we can content ourselves with the following statement, which holds in Euclidean spaces.

Proposition 3.1.8. *Let K be a **closed** subset of $\mathbb{R}^n$. Then, given any $x \in \mathbb{R}^n$, there exists a point $y \in K$ nearest to x, that is, such that*

$$\|x - y\| \leq \|x - z\| \quad \forall z \in K.$$

Proof. Given $x \in \mathbb{R}^n$, put

$$\delta = \inf_{z \in K} \|x - z\|. \tag{3.1.16}$$

It is a matter of proving that the infimum defined in (3.1.16) is actually a minimum, that is, it is attained at some point $y \in K$. At any rate, there exists a sequence $(z_n) \subset K$ such that

$$\|x - z_n\| \to \delta \quad (n \to \infty). \tag{3.1.17}$$

The sequence (z_n) is necessarily bounded (for otherwise, there would be a subsequence, still denoted (z_n) for convenience, such that $\|z_n\| \to \infty$ and therefore also $\|x - z_n\| \to \infty$, contradicting (3.1.17)). Therefore, by the Bolzano–Weierstrass theorem, (z_n) contains a convergent subsequence, which we call (z_{n_k}); and if we put

$$z_0 = \lim_{k\to\infty} z_{n_k},$$

then we have:
- $z_0 \in K$ (because K is closed);
- $\|x - z_0\| = \delta$ (because of (3.1.17) and the continuity of the norm).

Thus, z_0 has the required properties, and this ends the proof of Proposition 3.1.8. □

Exercise 3.1.2. Let X be a metric space and let $K \subset X$. Prove that if K is closed, then for any convergent sequence $(x_n) \subset K$ we have

$$\lim_{n\to\infty} x_n \in K. \tag{3.1.18}$$

Vice versa, if (3.1.18) holds whenever $(x_n) \subset K$ and (x_n) converges, then K is closed.

Exercise 3.1.3. Let E be a vector space with a norm denoted $\|\cdot\|$. Prove that the map

$$\|\cdot\| : E \to \mathbb{R}$$

is continuous.

Exercise 3.1.4. Let A be any set and let $f : A \to \mathbb{R}$ be bounded from below. Prove that there exists a sequence $(x_n) \subset A$ such that

$$\lim_{n\to\infty} f(x_n) = c \equiv \inf_{x\in A} f(x),$$

a similar property holding if f is bounded from above, with $C \equiv \sup_{x\in A} f(x)$.

3.2 Fourier series

A function $f : \mathbb{R} \to \mathbb{R}$ is said to be *periodic* of period $T > 0$ (or briefly, T-periodic) if

$$f(x + T) = f(x) \quad \forall x \in \mathbb{R}.$$

For instance, the trigonometric functions

$$\sin x, \cos x, \sin 2x, \cos 2x, \ldots \sin nx, \cos nx, \ldots, \tag{3.2.1}$$

where $n \in \mathbb{N}$, are all 2π-periodic.

Question. If we take an $f : \mathbb{R} \to \mathbb{R}$ that is 2π-periodic and – for instance – continuous, can we **expand f in series** of the trigonometric functions (3.2.1)? That is, does the equality

$$f(x) = a_0 + \sum_{k=1}^{\infty} (a_k \cos kx + b_k \sin kx) \tag{3.2.2}$$

hold for a suitable choice of the real constants a_0, a_k, b_k $(k \in \mathbb{N})$?

Definition 3.2.1. Given $f : \mathbb{R} \to \mathbb{R}$, which is 2π-periodic and continuous, the numbers

$$a_0 = \frac{1}{2\pi} \int_0^{2\pi} f(x)\,dx, \quad a_k = \frac{1}{\pi} \int_0^{2\pi} f(x) \cos kx\,dx, \quad b_k = \frac{1}{\pi} \int_0^{2\pi} f(x) \sin kx\,dx, \tag{3.2.3}$$

where $k \in \mathbb{N}$, are called the *Fourier coefficients* of f. The series (3.2.2) with the coefficients given by (3.2.3) is called the *Fourier series* of f.

Our question now becomes: given f, does its Fourier series converge to f?

We meet a similar question if we wish to expand f into **power series** near a point $x_0 \in \mathbb{R}$: in this case one writes down (provided f is of class C^∞) the **Taylor series** of f centered at x_0 and asks – taking, for instance, $x_0 = 0$ – if

$$f(x) = \sum_{k=0}^{\infty} a_k x^k, \quad \text{where } a_k = \frac{f^{(k)}(0)}{k!}.$$

One main point to discuss is **in what sense** the convergence of the Fourier series to f (that is, the equality in (3.2.2)) takes place. The most elementary meaning would be that for each fixed $x \in \mathbb{R}$, the numerical series in (3.2.2) converges precisely to the value $f(x)$ of f at x; this is **pointwise convergence**. However, this turns out to be quite delicate to prove and requires further assumptions on f, as can be seen in Section 4.3 of Chapter 4.

A stronger type of convergence would be **uniform** convergence (Chapter 2, Section 2.1).

One more type of convergence, which is particularly useful when dealing with Fourier series, is **convergence in the mean** – more precisely, in the **quadratic** mean, or mean of order 2, already introduced in Remark 3.1.2 and repeated here for the reader's convenience.

Definition 3.2.2. Let (f_n) be a sequence in $C([a,b])$. We say that (f_n) **converges in the mean to $f \in C([a,b])$** on the interval $[a,b]$ if

$$\int_a^b |f_n(x) - f(x)|^2\, dx \to 0 \quad \text{as } n \to \infty.$$

Remark 3.2.1. The definition above makes sense also in the more general case that the functions f_n $(n \in \mathbb{N})$ and f are merely **integrable** on $[a, b]$.

Example 3.2.1. Let $f_n(x) = x^n$ $(n \in \mathbb{N})$. We claim that (f_n) converges in the mean to $f = 0$ on the interval $[0, 1]$. Indeed,

$$\int_a^b |f_n(x) - f(x)|^2\, dx = \int_0^1 x^{2n}\, dx = \left[\frac{x^{(2n+1)}}{2n+1}\right]_0^1 = \frac{1}{2n+1} \to 0.$$

Note that the sequence (x^n) does not converge pointwisely to 0 on $[0, 1]$, for $\lim_{n\to\infty} x^n = 1$ if $x = 1$. In fact, in a sense, convergence in the mean is independent from pointwise convergence, as shown also by the next exercise.

Exercise 3.2.1. Let $f_n(x) = \sqrt{n^2 x e^{-nx}}$ $(n \in \mathbb{N})$. We clearly have

$$\lim_{n\to\infty} f_n(x) = 0 \quad \forall x \in [0, 1].$$

However, (f_n) does not converge in the mean to 0 on the interval $[0, 1]$, for

$$\lim_{n\to\infty} \int_0^1 n^2 x e^{-nx}\, dx \neq 0.$$

Remark 3.2.2. Definition 3.2.2 is useful and important also in other fields related to mathematical analysis, such as numerical analysis and probability theory. It can be generalized as follows: given a sequence (f_n) in $C([a, b])$ and given any $p \in \mathbb{R}$ with $p \geq 1$, we say that (f_n) *converges in the mean **of order** p* to $f \in C([a, b])$ on the interval $[a, b]$ if

$$\int_a^b |f_n(x) - f(x)|^p\, dx \to 0 \quad \text{as } n \to \infty.$$

One main result about the convergence of Fourier series that can be simply stated is the following.

Theorem 3.2.1. *Let $f : \mathbb{R} \to \mathbb{R}$ be 2π-periodic and continuous. Then its Fourier series converges in the mean to f on $[0, 2\pi]$; that is,*

$$\int_0^{2\pi} |s_n(x) - f(x)|^2\, dx \to 0 \quad \text{as } n \to \infty,$$

where

$$s_n(x) = a_0 + \sum_{k=1}^{n} (a_k \cos kx + b_k \sin kx) \quad (n \in \mathbb{N}),$$

with a_0, a_k, b_k $(k \in \mathbb{N})$ *as in* (3.2.3).

However, the proof of Theorem 3.2.1 is far from simple. One of its main ingredients will be the following relations, holding for $n, m \in \mathbb{N}$:

$$\int_0^{2\pi} \sin nx\, dx = \int_0^{2\pi} \cos nx\, dx = 0, \tag{3.2.4}$$

$$\int_0^{2\pi} \sin^2 nx\, dx = \int_0^{2\pi} \cos^2 nx\, dx = \pi, \tag{3.2.5}$$

$$\int_0^{2\pi} \sin nx \sin mx\, dx = \int_0^{2\pi} \cos nx \cos mx\, dx = 0 \quad (n \neq m), \tag{3.2.6}$$

$$\int_0^{2\pi} \sin nx \cos mx\, dx = 0. \tag{3.2.7}$$

Exercise 3.2.2. To prove the **orthogonality relations** (3.2.4)–(3.2.7), first recall that

$$\begin{cases} \cos^2 x + \sin^2 x = 1 \\ \cos^2 x - \sin^2 x = \cos 2x, \end{cases}$$

whence

$$2\cos^2 x = 1 + \cos 2x, \quad 2\sin^2 x = 1 - \cos 2x, \tag{3.2.8}$$

so that

$$\int \cos^2 x\, dx = \frac{1}{2}\left[x + \frac{\sin 2x}{2}\right],$$
$$\int \sin^2 x\, dx = \frac{1}{2}\left[x - \frac{\sin 2x}{2}\right].$$

From these, integrating from 0 to 2π yields (3.2.5).

Note that (3.2.8) are special cases of the following relations:

$$\begin{cases} 2\cos\alpha\cos\beta = \cos(\alpha - \beta) + \cos(\alpha + \beta) \\ 2\sin\alpha\sin\beta = \cos(\alpha - \beta) - \cos(\alpha + \beta) \\ 2\sin\alpha\cos\beta = \sin(\alpha - \beta) + \sin(\alpha + \beta), \end{cases} \tag{3.2.9}$$

which in turn follow from the formulae for $\sin(\alpha + \beta)$ and $\cos(\alpha + \beta)$.

From (3.2.9) we obtain, for instance, the equality

$$\int \sin nx \sin mx \, dx = \frac{1}{2} \int \cos(n-m)x \, dx - \frac{1}{2} \int \cos(n+m)x \, dx,$$

which explains the first of (3.2.6). The second is obtained similarly. Finally, the last equation in (3.2.9) yields (3.2.7).

Definition 3.2.3. A **trigonometric polynomial** *of degree* $\leq n$ $(n \in \mathbb{N})$ is an expression of the form

$$P(x) = a_0 + \sum_{k=1}^{n} (a_k \cos kx + b_k \sin kx), \tag{3.2.10}$$

where $a_0, a_k, b_k \in \mathbb{R}$. The polynomial (3.2.10) is said to be *of degree n* if a_n and/or b_n are different from 0.

Fix an integer $n \in \mathbb{N}$ and define

$$Q_n \equiv \{P \mid P \text{ is a trigonometric polynomial of degree} \leq n\}. \tag{3.2.11}$$

We note the following:

- Q_n is a vector subspace of $C_{2\pi}(\mathbb{R})$, where

$$C_{2\pi}(\mathbb{R}) \equiv \{f : \mathbb{R} \to \mathbb{R} \mid f \text{ is continuous and } 2\pi\text{-periodic}\}.$$

- Precisely, Q_n is the vector subspace spanned by the functions

$$1, \cos x, \sin x, \cos 2x, \sin 2x, \ldots, \cos nx, \sin nx. \tag{3.2.12}$$

- Using the orthogonality relations (3.2.4)–(3.2.7), it follows that the functions above are linearly independent, and therefore $\dim Q_n = 2n + 1$.

Given $f \in C_{2\pi}(\mathbb{R})$, the nth partial sum S_n of its Fourier series,

$$S_n(x) = a_0 + \sum_{k=1}^{n} (a_k \cos kx + b_k \sin kx) \quad (n \in \mathbb{N}), \tag{3.2.13}$$

where a_0, a_k, b_k are as in (3.2.3), belongs to Q_n.

For the proof of Theorem 3.2.1, one key step will be the following fact.

Proposition 3.2.1. *Let $f \in C_{2\pi}(\mathbb{R})$ be given. Then among all trigonometric polynomials of degree $\leq n$, S_n* ***is the nearest to*** *f; here “near” means in the sense of the* ***distance given by the quadratic mean***

$$d_2(f,g) = \|f - g\|_2 = \left(\int_0^{2\pi} |f(x) - g(x)|^2 \, dx \right)^{1/2}.$$

*In other words, S_n is the **best approximation** to f among all vectors of Q_n:*

$$\|f - S_n\|_2 \leq \|f - P\|_2 \quad \forall P \in Q_n. \tag{3.2.14}$$

Enlarging the space of 2π-periodic functions: from $C_{2\pi}(\mathbb{R})$ to $\hat{C}_{2\pi}(\mathbb{R})$

The requirement of continuity is unnecessarily strong for the study of Fourier series. Here we consider some extensions, first considering functions defined in a bounded closed interval $[a,b]$ and then moving on to functions defined on the whole of $\mathbb{R}$.

Piecewise continuous functions

Definition 3.2.4. A function $f : [a,b] \to \mathbb{R}$ is said to be **piecewise continuous** in $[a,b]$ if it is continuous at every point of $[a,b]$ except at most in a finite number of them, call them $x_1, \ldots, x_n$, in which however there exist and are finite the left and right limits

$$f(x_i^-) \equiv \lim_{x \to x_i^-} f(x), \quad f(x_i^+) \equiv \lim_{x \to x_i^+} f(x)$$

if $a < x_i < b$; we have a similar requirement if $x_1 = a$ or $x_n = b$.

Figures 3.1 and 3.2 help to visualize the character of piecewise continuous functions.

Remark 3.2.3. If $f : [a,b] \to \mathbb{R}$ is piecewise continuous, then f is (i) bounded and (ii) integrable on $[a,b]$.

Example 3.2.2. A very simple – though very useful in approximation theory – class of piecewise continuous functions is that of **piecewise linear** functions; these are the special f's whose restriction to each of the intervals $[x_i, x_{i+1}]$ is linear (see Figure 3.3):

$$f(x) = a_i x + b_i \quad (x_i < x < x_{i+1}).$$

We have in this case

$$f(x_i^-) = a_{i-1} x_i + b_{i-1}, \quad f(x_i^+) = a_i x_i + b_i.$$

Definition 3.2.5. A piecewise continuous function $f : [a,b] \to \mathbb{R}$ is said to be **regular** if for every point $x_0 \in\,]a,b[$,

$$f(x_0) = \frac{f(x_0^+) + f(x_0^-)}{2}. \tag{3.2.15}$$

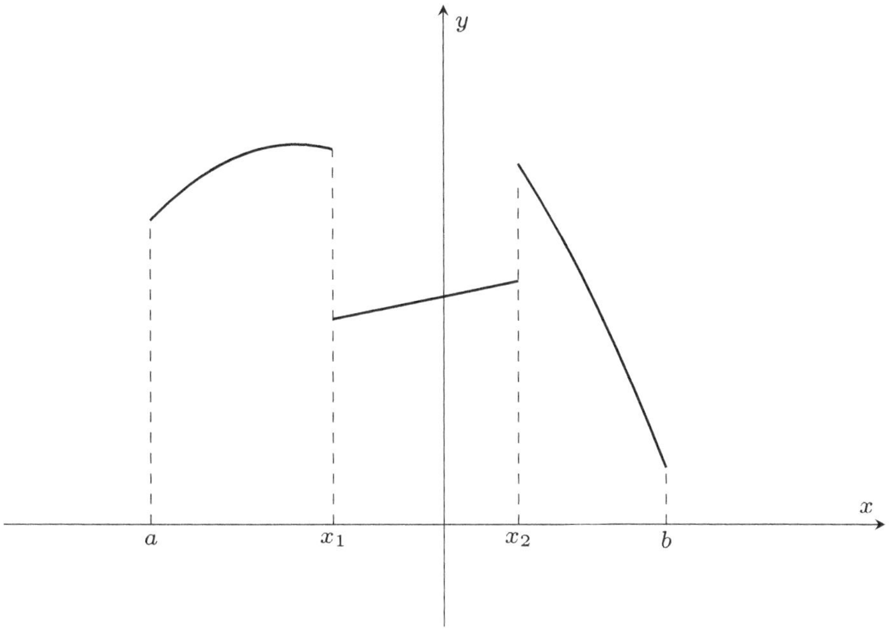

Figure 3.1: A piecewise continuous f.

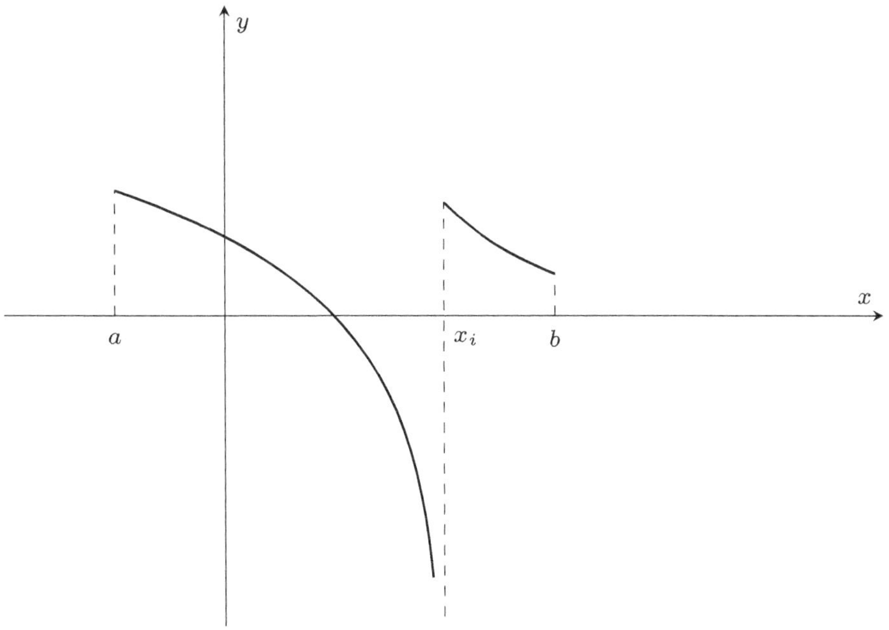

Figure 3.2: An f that is not piecewise continuous.

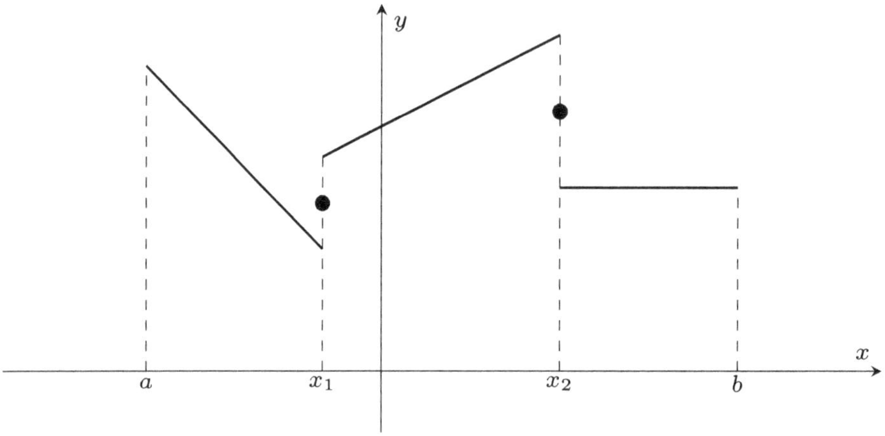

Figure 3.3: A piecewise linear f.

It is clear from the above definitions that:
- If f is continuous in $[a, b]$, then it is piecewise continuous and regular.
- If f is merely piecewise continuous in $[a, b]$, then it can always be **regularized** – that is, made regular – just changing its values in the (finitely many) points of discontinuity x_i, that is, putting **by definition** $f(x_i)$ equal to the mean value of $f(x_i^-), f(x_i^+)$ as prescribed by (3.2.15).

Back to the 2π-periodic functions

Definition 3.2.6. A function $f : \mathbb{R} \to \mathbb{R}$ is said to be *piecewise continuous* (resp. *regular*) if it is piecewise continuous (resp. regular) on every closed bounded interval $[a, b]$.

Here and henceforth we put

$$\hat{C}_{2\pi}(\mathbb{R}) \equiv \{f : \mathbb{R} \to \mathbb{R} : f \text{ is } 2\pi\text{-periodic, piecewise continuous, and regular}\}$$

Of course, the trigonometric functions in (3.2.1) are the most familiar representatives of $\hat{C}_{2\pi}(\mathbb{R})$. Figure 3.4 gives an idea of how to create many new examples of functions belonging to this space.

More formally, we have the following.

Example 3.2.3. Let $f : [a, a+2\pi] \to \mathbb{R}$. If $f \in \hat{C}[a, a+2\pi]$, then its 2π**-periodic extension** $\hat{f}$ defined putting

$$\hat{f}(x + 2k\pi) = f(x), \quad x \in [a, a + 2\pi[,$$

belongs to $\hat{C}_{2\pi}(\mathbb{R})$.

A few specific examples now follow.

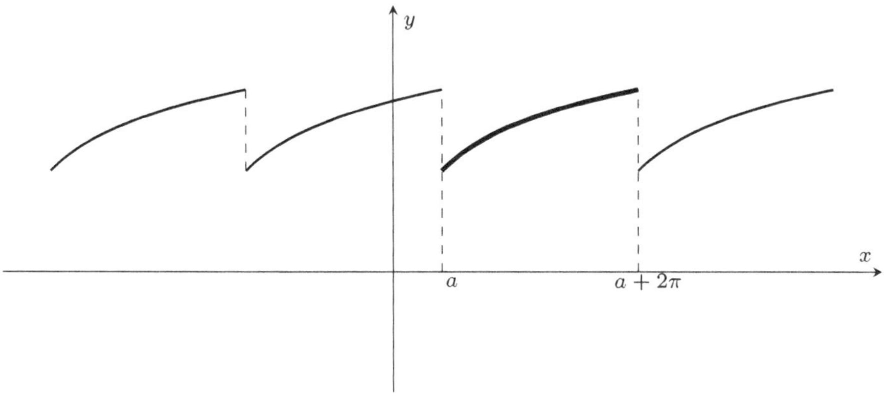

Figure 3.4: The periodic extension $\hat{f}$ of f.

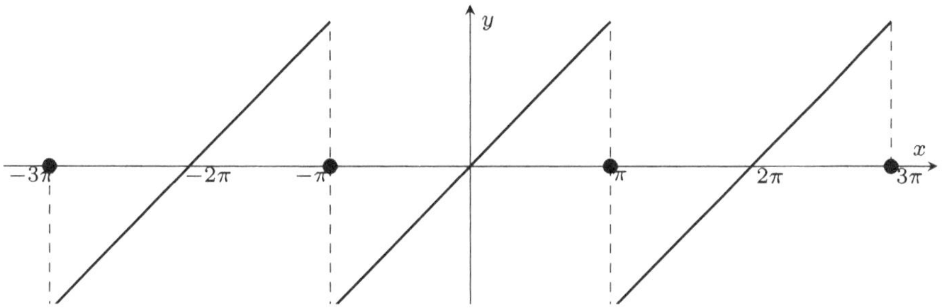

Figure 3.5: $f(x) = x, -\pi \le x \le \pi$.

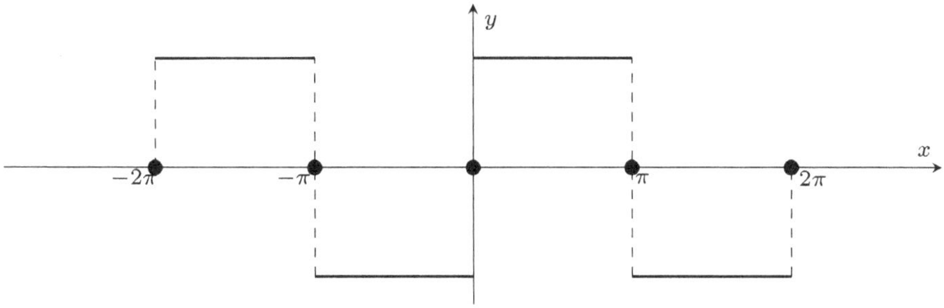

Figure 3.6: $f(x) = -1\,(-\pi < x < 0); f(x) = 1\,(0 < x < \pi)$.

Example 3.2.4. Consider first the 2π-periodic extension of $f(x) = x, |x| \le \pi$ (see Figure 3.5).

Example 3.2.5. Consider now the 2π-periodic extension of the so-called *Heaviside function*: $f(x) = -1\,(-\pi < x < 0); f(x) = 1\,(0 < x < \pi)$; see Figure 3.6.

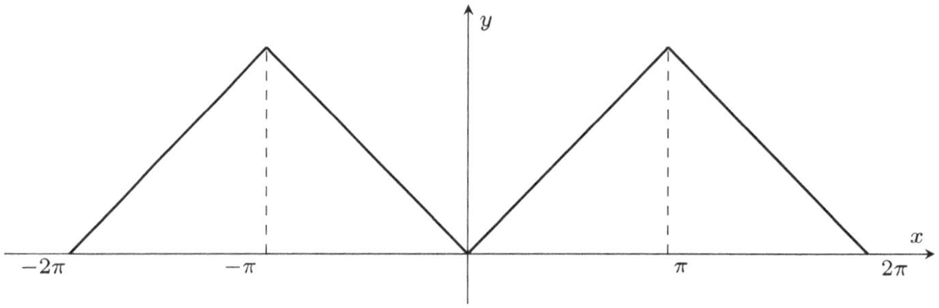

Figure 3.7: $f(x) = |x|, -\pi \le x \le \pi$.

Example 3.2.6. Consider finally $\hat{f}$ when f is the absolute value function (again for $|x| \le \pi$): $f(x) = |x|, -\pi \le x \le \pi$; see Figure 3.7.

Exercise 3.2.3. For each of Examples 3.2.4, 3.2.5, and 3.2.6 compute the respective Fourier series. The computations can be easily performed and give respectively the following results:

$$a_0 = a_k = 0, \quad b_k = (-1)^{k+1}\frac{2}{k} \quad (k \in \mathbb{N}),$$
$$a_0 = a_k = 0, \quad b_k = \frac{2}{k\pi}[1 - (-1)^k] \quad (k \in \mathbb{N}),$$
$$a_0 = \frac{\pi}{2}, \quad a_k = -\frac{2}{k^2\pi}[1 - (-1)^k], \quad b_k = 0 \quad (k \in \mathbb{N}).$$

In order to handle Exercise 3.2.3 or similar – and in general, in order to compute the Fourier coefficients (3.2.3) of a given function – the following three remarks turn out to be useful.

Remark 3.2.4. If $f \in \hat{C}_{2\pi}(\mathbb{R})$, then

$$\int_0^{2\pi} f(x)\,dx = \int_a^{a+2\pi} f(x)\,dx \quad \forall a \in \mathbb{R}.$$

Indeed,

$$\int_a^{a+2\pi} f(x)\,dx = \int_a^0 + \int_0^{2\pi} + \int_{2\pi}^{a+2\pi}. \tag{3.2.16}$$

In the last integral, make the change of variable $x = y + 2\pi$ and use the 2π-periodicity of f to obtain

$$\int_{2\pi}^{a+2\pi} f(x)\,dx = \int_0^a f(y + 2\pi)\,dx = \int_0^a f(y)\,dy$$

so that the last integral in (3.2.16) will cancel with the first.

Remark 3.2.5. Modifying a piecewise continuous $f : [a,b] \to \mathbb{R}$ in a finite number of points does not change its integral on $[a,b]$. In particular, in the computation of the Fourier coefficients a_0, a_k, b_k we need not worry if f has been regularized or not.

Remark 3.2.6. Let $f : [-a,a] \to \mathbb{R}$ be piecewise continuous. Then if f is *odd* (that is, $f(-x) = -f(x)$), we have

$$\int_{-a}^{a} f(x)\,dx = 0,$$

while if f is *even* (that is, $f(-x) = f(x)$), we have

$$\int_{-a}^{a} f(x)\,dx = 2\int_{0}^{a} f(x)\,dx.$$

Now since – by Remark 3.2.4 – we can compute the integrals defining the Fourier coefficients a_0, a_k, b_k of an $f \in \hat{C}_{2\pi}(\mathbb{R})$ integrating on $[-\pi,\pi]$ rather than on $[0,2\pi]$, it follows in particular that if f is odd, then

$$\begin{cases} a_0 = \frac{1}{2\pi}\int_{-\pi}^{\pi} f(x)\,dx = 0, \quad a_k = \frac{1}{\pi}\int_{-\pi}^{\pi} f(x)\cos kx\,dx = 0 \\ b_k = \frac{1}{\pi}\int_{-\pi}^{\pi} f(x)\sin kx\,dx = \frac{2}{\pi}\int_{0}^{\pi} f(x)\sin kx\,dx \end{cases} \tag{3.2.17}$$

so that the Fourier series of f will contain only sine terms, that is, it will be a **Fourier sine series**, while if f is even, then it will be developed into a **Fourier cosine series**.

Inner product for piecewise continuous functions

Given an interval $[a,b]$, put

$$\hat{C}[a,b] \equiv \{f : [a,b] \to \mathbb{R} : f \text{ is piecewise continuous and regular}\}.$$

Evidently, $\hat{C}[a,b]$ is a vector space. Moreover, we can extend to this vector space the scalar product defined in $C[a,b]$ (Example 3.1.2), for we have the following proposition.

Proposition 3.2.2. *Given $f, g \in \hat{C}[a,b]$, put*

$$\langle f, g\rangle \equiv \int_{a}^{b} f(x)g(x)\,dx. \tag{3.2.18}$$

Then (3.2.18) *defines an inner product in $\hat{C}[a,b]$.*

In order to verify the positive definiteness of the inner product (3.2.18), the following extension of Proposition 3.1.1 is useful.

Proposition 3.2.3. *Let $g : [a,b] \to \mathbb{R}$ be continuous in $]a,b[$ and have finite one-sided limits $g(a^+), g(b^-)$ at the endpoints a and b. If $g \geq 0$ in $[a,b]$ and $\int_a^b g(x)dx = 0$, then $g(x) = 0$ for every $x \in]a,b[$, so that also $g(a^+), g(b^-) = 0$.*

In turn, we can extend Proposition 3.2.3 to a **piecewise continuous regular function** g defined in $[a,b]$ and such that $g \geq 0$ in $[a,b]$ and $\int_a^b g(x)dx = 0$. Indeed, let $a \leq x_1 \leq x_2 \cdots \leq x_n \leq b$ be the discontinuity points of g, in which by assumption there exist the finite left and right limits $g(x_i^-), g(x_i^+)$. We have (putting for convenience $x_0 \equiv a, x_{n+1} \equiv b$)

$$\int_a^b g(x)dx = \sum_{i=0}^{n} \int_{x_i}^{x_{i+1}} g(x)\,dx,$$

and by our assumptions, each term in the sum of the right-hand side is equal to zero. Applying Proposition 3.2.3 to each of the intervals $[x_i, x_{i+1}]$, it follows that $g = 0$ in each of the open intervals $]x_i, x_{i+1}[$, and therefore necessarily $g(x_i^-) = g(x_i^+) = 0$ for every $i = 1, \ldots, n$. Since g is regular, it then follows that $g(x_i) = 0$ for every i, so that $g \equiv 0$ in $[a,b]$.

Given $f \in \hat{C}[a,b]$, the remarks above explain the implication $\langle f, f\rangle = 0 \Rightarrow f = 0$ and thus clarify and complete the statement of Proposition 3.2.2.

We conclude these technical but necessary remarks with an explicit statement that is just the “2π-periodic version” of Proposition 3.2.2.

Proposition 3.2.4. *Given $f, g \in \hat{C}_{2\pi}(\mathbb{R})$, put*

$$\langle f, g\rangle \equiv \int_0^{2\pi} f(x)g(x)\,dx. \tag{3.2.19}$$

Then (3.2.19) *defines an inner product in $\hat{C}_{2\pi}(\mathbb{R})$.*

Back to Fourier series

Recall the orthogonality relations for the trigonometric functions:

$$\int_0^{2\pi} \sin nx\,dx = \int_0^{2\pi} \cos nx\,dx = 0, \tag{3.2.20}$$

$$\int_0^{2\pi} \sin^2 nx\,dx = \int_0^{2\pi} \cos^2 nx\,dx = \pi, \tag{3.2.21}$$

$$\int_0^{2\pi} \sin nx \sin mx\, dx = \int_0^{2\pi} \cos nx \cos mx\, dx = 0 \quad (n \neq m), \tag{3.2.22}$$

$$\int_0^{2\pi} \sin nx \cos mx\, dx = 0. \tag{3.2.23}$$

We use the following notations:

$$e_0 = \frac{1}{\sqrt{2\pi}}, \quad e_1(x) = \frac{\cos x}{\sqrt{\pi}}, \quad e_2(x) = \frac{\sin x}{\sqrt{\pi}}, \quad e_3(x) = \frac{\cos 2x}{\sqrt{\pi}}, \quad e_4(x) = \frac{\sin 2x}{\sqrt{\pi}}, \dots, \tag{3.2.24}$$

that is,

$$e_0 = \frac{1}{\sqrt{2\pi}}, \quad e_{2k-1}(x) = \frac{\cos kx}{\sqrt{\pi}}, \quad e_{2k}(x) = \frac{\sin kx}{\sqrt{\pi}} \quad (k \in \mathbb{N}). \tag{3.2.25}$$

With these notations, the orthogonality relations (3.2.20)–(3.2.23) can be resumed in the following statement.

Proposition 3.2.5. *The family* $\{e_n \mid n = 0, 1, \dots\}$ *defined in* (3.2.25) *forms an orthonormal sequence in* $\hat{C}_{2\pi}(\mathbb{R})$.

Proof. First, we have

$$\|e_n\| = 1 \quad \forall k = 0, 1, 2, \dots.$$

This is evident for $n = 0$, while for $n \in \mathbb{N}$ we use (3.2.21) to see that if, e. g., $n = 2k - 1$ is an odd integer,

$$\|e_n\|^2 = \frac{1}{\pi} \int_0^{2\pi} \cos^2 kx\, dx = \frac{1}{\pi}\pi = 1.$$

Moreover, we have

$$\langle e_n, e_m \rangle = 0 \quad \forall n \neq m. \tag{3.2.26}$$

Indeed, if $n = 0$ and $m \in \mathbb{N}$ we use (3.2.20) to see that if, e. g., $m = 2k$ is even,

$$\langle e_0, e_m \rangle = \int_0^{2\pi} \frac{1}{\sqrt{2\pi}} \frac{\sin kx}{\sqrt{\pi}}\, dx = 0.$$

If $n, m \geq 1$, then (3.2.26) follows from (3.2.22) and (3.2.23). □

We can now use the general results about inner product spaces discussed in Section 3.1 of this chapter, in particular Theorem 3.1.1. We have indeed the following proposition.

Proposition 3.2.6. *Given $f \in \hat{C}_{2\pi}(\mathbb{R})$, the nth partial sum S_n of its Fourier series is the* ***orthogonal projection of f on the subspace*** *Q_n defined in* (3.2.11), *that is,*

$$Q_n \equiv \{P \mid P \textit{ is a trigonometric polynomial of degree} \leq n\}.$$

Proof. Let $P_n(f)$ denote the orthogonal projection of f onto Q_n. By Proposition 3.2.5 and the definition of Q_n,

$$e_0, e_1, e_2, \dots, e_{2n-1}, e_{2n}$$

is an orthonormal basis for Q_n. Therefore, by formula (3.1.12) of Theorem 3.1.1 – in which we take $E = \hat{C}_{2\pi}(\mathbb{R})$ and $M = Q_n$ (n is fixed) – we have

$$P_n(f) = \sum_{i=0}^{2n} \langle f, e_i\rangle e_i = \langle f, e_0\rangle e_0 + \langle f, e_1\rangle e_1 + \cdots + \langle f, e_{2n}\rangle e_{2n}. \tag{3.2.27}$$

Recall the definition (3.2.3) of the Fourier coefficients of f:

$$a_0 = \frac{1}{2\pi}\int_0^{2\pi} f(x)\,dx, \quad a_k = \frac{1}{\pi}\int_0^{2\pi} f(x)\cos kx\,dx, \quad b_k = \frac{1}{\pi}\int_0^{2\pi} f(x)\sin kx\,dx.$$

Looking at these and at the definition (3.2.25) of e_i, we see that

$$\langle f, e_0\rangle = \sqrt{2\pi}a_0, \quad \langle f, e_i\rangle = \begin{cases} \sqrt{\pi}a_k, & i = 2k-1 \\ \sqrt{\pi}b_k, & i = 2k. \end{cases} \tag{3.2.28}$$

Using these equalities in (3.2.27), we then have

$$\begin{aligned} P_n(f) &= \sqrt{2\pi}a_0\frac{1}{\sqrt{2\pi}} + \sqrt{\pi}a_1\frac{\cos x}{\sqrt{\pi}} + \cdots + \sqrt{\pi}b_n\frac{\sin nx}{\sqrt{\pi}} \\ &= a_0 + \sum_{k=1}^{n}(a_k\cos kx + b_k\sin kx) = S_n. \end{aligned}$$

The proposition above implies, again by Theorem 3.1.1, the best approximation property indicated for S_n by Proposition 3.2.1, which we report here practically unaltered, save replacing $C_{2\pi}(\mathbb{R})$ with the larger space $\hat{C}_{2\pi}(\mathbb{R})$. □

Proposition 3.2.7. *Let $f \in \hat{C}_{2\pi}(\mathbb{R})$ be given. Then among all trigonometric polynomials of degree $\leq n$, S_n* ***is the nearest to*** *f in the sense of the* ***distance given by the quadratic mean***

$$d_2(f,g) = \|f - g\|_2 = \left(\int_0^{2\pi} |f(x) - g(x)|^2 \, dx \right)^{1/2}.$$

In other words, S_n is the **best approximation** *to f among all vectors of Q_n:*

$$\|f - S_n\|_2 \le \|f - P\|_2 \quad \forall P \in Q_n. \tag{3.2.29}$$

Proof. Take $E = \hat{C}_{2\pi}(\mathbb{R})$, $M = Q_n$ and use formula (3.1.13) in Theorem 3.1.1, exploiting the characterization of S_n given by Proposition 3.2.6 above.

We are now in a position to prove Theorem 3.2.1, and in fact its more general version extended to piecewise continuous functions. Here is the formal statement. □

Theorem 3.2.2. *Let $f \in \hat{C}_{2\pi}(\mathbb{R})$. Then its Fourier series converges* **in the quadratic mean** *to f on $[0, 2\pi]$; that is,*

$$\lim_{n\to\infty} \int_0^{2\pi} |S_n(x) - f(x)|^2 \, dx = 0, \tag{3.2.30}$$

where

$$S_n(x) = a_0 + \sum_{k=1}^{n} (a_k \cos kx + b_k \sin kx) \quad (n \in \mathbb{N})$$

(with a_0, a_k, b_k as in (3.2.3)) is the nth partial sum of the Fourier series of f.

The proof of Theorem 3.2.2 rests on the characterization of S_n just found in Proposition 3.2.7 **and** on the following two lemmas about the approximation of functions: the first is related to the Weierstrass theorem on the **uniform** approximation of a continuous function by polynomials, while the second allows to approximate **in the mean** a piecewise continuous function with a continuous one.

Lemma 3.2.1. *Let $f \in C_{2\pi}(\mathbb{R})$. Then there exists a sequence $(T_n)_{n\in\mathbb{N}}$ of trigonometric polynomials, with $T_n \in Q_n$ $\forall n \in \mathbb{N}$, such that $T_n \to f$* uniformly *on $[0, 2\pi]$, that is,*

$$\lim_{n\to\infty} \|T_n - f\|_\infty = \lim_{n\to\infty} \left(\sup_{x\in[0,2\pi]} |T_n(x) - f(x)| \right) = 0. \tag{3.2.31}$$

Remark 3.2.7. Lemma 3.2.1 is an important result on **approximation by polynomials**: it can be viewed as the "periodic version" of the famous **Weierstrass approximation theorem**, stating that any continuous function on an interval $[a, b]$ is the limit of a sequence of polynomials which converges uniformly in $[a, b]$. For a proof of these results, the interested reader can look into W. Rudin's books [10] and [6]; see also Chapter VII, Section 4 of Dieudonné [7].

Remark 3.2.8. The uniform convergence of a sequence of continuous (or, more generally, integrable) functions on an interval $[a, b]$ implies its convergence in the mean of any order p; indeed, we have

$$\int_a^b |f_n(x) - f(x)|^p\,dx \leq \int_a^b \Big(\sup_{x\in[a,b]} |f_n(x) - f(x)|\Big)^p\,dx = \|f_n - f\|_\infty^p (b-a).$$

Lemma 3.2.2. *Let $f \in \hat{C}_{2\pi}(\mathbb{R})$. Then given any $\epsilon > 0$, there exists a $g \in C_{2\pi}(\mathbb{R})$ such that*

$$\|f - g\|_2 < \epsilon. \tag{3.2.32}$$

Proof. Let $f \in \hat{C}_{2\pi}(\mathbb{R})$. By definition, f has at most a finite number of points of discontinuity $x_1, \ldots, x_n$ in the interval $[0, 2\pi]$, and the one-sided limits at these points exist and are finite. Suppose for simplicity that f has only one such discontinuity point x_1 and that $x_1 \in]0, 2\pi[$.

Fix $\epsilon > 0$. Let $\delta > 0$ be such that $[x_1 - \delta, x_1 + \delta] \subset]0, 2\pi[$ and define g_δ in $[0, 2\pi]$ as follows (see Figure 3.8):

$$g_\delta(x) = \begin{cases} f(x), & |x - x_1| > \delta \\ ax + b, & |x - x_1| \leq \delta, \end{cases} \tag{3.2.33}$$

where a, b are chosen so that g_δ is **continuous** in $[0, 2\pi]$, that is, a and b are determined by the equations

$$\begin{cases} a(x_1 - \delta) + b = f(x_1 - \delta) \\ a(x_1 + \delta) + b = f(x_1 + \delta). \end{cases} \tag{3.2.34}$$

Let g be the 2π-periodic extension of g_δ to all of $\mathbb{R}$. Then (since $g_\delta(0) = g_\delta(2\pi)$) $g \in C_{2\pi}(\mathbb{R})$. Also note that by construction, $|g_\delta(x)| \leq K \equiv \sup_{x\in[0,2\pi]} |f(x)|$; therefore, we have

$$\|f - g\|_2^2 = \int_0^{2\pi} |f(x) - g_\delta(x)|^2\,dx = \int_{x_1-\delta}^{x_1+\delta} |f(x) - g_\delta(x)|^2\,dx \leq 8\delta K^2$$

so that (3.2.32) will be satisfied as soon as δ is taken so small that $8\delta K^2 < \epsilon^2$.

We can now **conclude the proof of Theorem 3.2.2.**

We need to prove that given any $f \in \hat{C}_{2\pi}(\mathbb{R})$, one has $\|S_n - f\|_2 \to 0$ as $n \to \infty$, where S_n is the nth partial sum of the Fourier series of f; that is, given any $\epsilon > 0$, we need to find an $\hat{n} \in \mathbb{N}$ such that

$$\|S_n - f\|_2 < \epsilon \quad \forall n \in \mathbb{N}, n \geq \hat{n}. \tag{3.2.35}$$

By virtue of Lemma 3.2.2, we first find a $g \in C_{2\pi}(\mathbb{R})$ such that

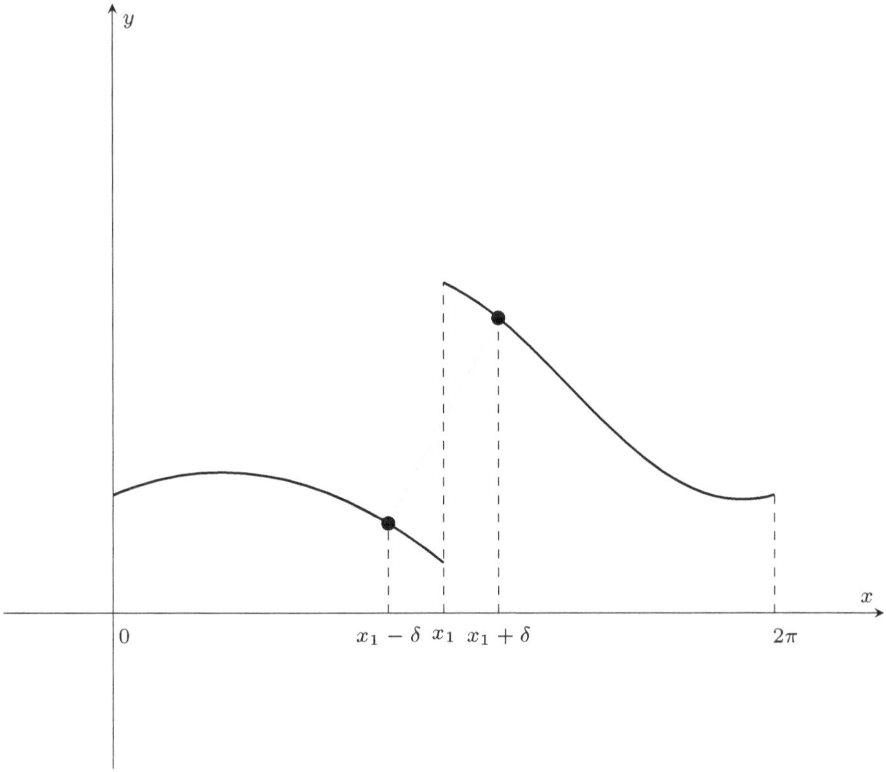

Figure 3.8: Correcting a discontinuity.

$$\|f - g\|_2 < \frac{\epsilon}{2}. \tag{3.2.36}$$

Next, on the basis of Lemma 3.2.1, there exists a sequence (T_n) of trigonometric polynomials, with $T_n \in Q_n \ \forall n \in \mathbb{N}$, converging uniformly (and therefore also in the quadratic mean, see Remark 3.2.8) to g; thus, we find an $n_0 \in \mathbb{N}$ such that

$$\|g - T_n\|_2 < \frac{\epsilon}{2} \quad \forall n \in \mathbb{N}, n \geq n_0. \tag{3.2.37}$$

Using (3.2.36) and (3.2.37) we thus have

$$\|f - T_n\|_2 < \epsilon \quad \forall n \in \mathbb{N}, n \geq n_0. \tag{3.2.38}$$

However, by Proposition 3.2.7 and in particular by formula (3.2.29) we have

$$\|f - S_n\|_2 \leq \|f - T_n\|_2 \quad \forall n \in \mathbb{N}$$

so that the conclusion (that is, the proof of our claim (3.2.35)) follows from (3.2.38) on taking $\hat{n} = n_0$. □

Remark 3.2.9. Beware that (3.2.38) does **not** mean that $\|f - T_n\|_2 \to 0$ as $n \to \infty$. Indeed, (T_n) depends upon g, and therefore upon ϵ!

Remark 3.2.10. Lemma 3.2.2 shows, more generally, that a piecewise continuous function can be approximated **in the mean** as closely as we wish by continuous functions. On the other hand, note that it is **not** possible to approximate **uniformly** a discontinuous function with continuous ones (see Proposition 2.3.3).

Remark 3.2.11. To better appreciate the meaning of the term "approximation" used in many statements of this section, it is useful to reason in general terms and carry out the following exercise. Let (X, d) be a metric space, let $A \subset X$, and let $x_0 \in X$. Then the following are equivalent:
(i) for any $\epsilon > 0$, there exists an $x \in A$ such that $d(x, x_0) < \epsilon$;
(ii) there exists a sequence $(x_n) \subset A$ such that $\lim_{n\to\infty} x_n = x_0$.

Such property of x_0 is expressed by saying that x_0 is a **cluster point of** A. The set of all cluster points of A is denoted with $\overline{A}$ and called the **closure of** A. These definitions and concepts were proposed also in Chapter 2, Section 2.2; see in particular Proposition 2.2.3.

The final part of this section is devoted to some important and useful consequences of the basic Theorem 3.2.2.

Corollary 3.2.1 (Parseval's identity). *Let $f \in \hat{C}_{2\pi}(\mathbb{R})$ and let a_0, a_k, b_k $(k \in \mathbb{N})$ be its Fourier coefficients as defined in* (3.2.3). *Then*

$$\int_0^{2\pi} f^2(x)\,dx = 2\pi a_0^2 + \pi \sum_{k=1}^{\infty} (a_k^2 + b_k^2). \tag{3.2.39}$$

Proof. We have seen in Theorem 3.1.1, and especially in formula (3.1.14), that if y is the orthogonal projection of x onto the subspace $M = [e_1, \dots, e_n]$, then

$$\|x - y\|^2 = \|x\|^2 - \sum_{i=1}^{n} \langle x, e_i\rangle^2. \tag{3.2.40}$$

In our case we have

$$\|f - S_n\|_2^2 = \|f\|_2^2 - \sum_{i=0}^{2n} \langle f, e_i\rangle^2. \tag{3.2.41}$$

Equality (3.2.41) holds for every $n \in \mathbb{N}$ (for we can project orthogonally f on every subspace in the increasing sequence $(Q_n)_{n\in\mathbb{N}}$). Moreover, from (3.2.28) we have

$$\langle f, e_0\rangle = \sqrt{2\pi}a_0, \quad \langle f, e_i\rangle = \begin{cases} \sqrt{\pi}a_k, & i = 2k-1 \\ \sqrt{\pi}b_k, & i = 2k \end{cases} \tag{3.2.42}$$

so that (3.2.41) yields

$$\|f - S_n\|_2^2 = \|f\|_2^2 - \left(2\pi a_0^2 + \pi \sum_{k=1}^{n} (a_k^2 + b_k^2)\right). \tag{3.2.43}$$

This first shows that

$$\left(2\pi a_0^2 + \pi \sum_{k=1}^{n} (a_k^2 + b_k^2)\right) \le \|f\|_2^2$$

for every $n \in \mathbb{N}$ and thus proves that the numerical series in (3.2.39) is convergent; moreover, writing (3.2.43) as

$$\|f\|_2^2 = \left(2\pi a_0^2 + \pi \sum_{k=1}^{n} (a_k^2 + b_k^2)\right) + \|f - S_n\|_2^2 \tag{3.2.44}$$

and letting $n \to \infty$, we obtain (3.2.39) as a consequence of (3.2.30) in Theorem 3.2.2. □

Corollary 3.2.2. *Let $f, g \in \hat{C}_{2\pi}(\mathbb{R})$. If f and g have the same Fourier coefficients, then they are equal.*

Proof. Let $h = f - g$. Then $h \in \hat{C}_{2\pi}(\mathbb{R})$ and

$$a_k(h) = \frac{1}{\pi} \int_0^{2\pi} (f(x) - g(x)) \cos kx \, dx = a_k(f) - a_k(g) = 0 \quad (k \in \mathbb{N})$$

and similarly for $a_0(h)$ and $b_k(h)$. It then follows from (3.2.39) that

$$\int_0^{2\pi} h^2(x) \, dx = 0$$

and therefore $h(x) = 0 \quad \forall x \in [0, 2\pi]$, whence the result. □

Corollary 3.2.3 (Riemann–Lebesgue's lemma). *Let $f \in \hat{C}_{2\pi}(\mathbb{R})$. Then*

$$\lim_{k\to\infty} \int_0^{2\pi} f(x) \cos kx \, dx = \lim_{k\to\infty} \int_0^{2\pi} f(x) \sin kx \, dx = 0. \tag{3.2.45}$$

Proof. This is an immediate consequence of the convergence of the series in (3.2.39); indeed, we have

$$0 \le a_k^2, b_k^2 \le a_k^2 + b_k^2 \to 0 \quad (k \to \infty),$$

whence

$$\lim_{k\to\infty} a_k = \lim_{k\to\infty} b_k = 0. \qquad \square$$

Further questions to discuss about Fourier series

1. Pointwise and uniform convergence of Fourier series

The convergence in the mean of the Fourier series of $f \in \hat{C}_{2\pi}(\mathbb{R})$ to f itself proved in Theorem 3.2.2 does not tell us anything about its pointwise convergence, neither a fortiori about its uniform convergence. As these forms of convergence will be useful in the construction of series solutions to some BVPs for PDEs, we shall resume these questions – under stronger regularity assumptions on f, in particular some differentiability property – in the next chapter (Section 4.3) when discussing PDEs.

2. Abstract (i. e., general) versions of the results discussed in this section

To a good extent, the results about the convergence in the mean of the Fourier series are due to the inner product space structure of the space $\hat{C}_{2\pi}(\mathbb{R})$, a leading role having been played by the orthonormality of the trigonometric system and by the concept of orthogonal projection onto a subspace. Could anything similar be established for general orthonormal systems in any inner product space? The next section is devoted precisely to answering this question. In the subsequent section (Section 3.4) we shall see what more can be said in the case of *complete* inner product spaces, that is, Hilbert spaces.

3.3 Orthonormal systems in inner product spaces. Sequence spaces

1. A useful equality

Let $e_1, \dots, e_n$ be orthonormal vectors of an inner product space E. Then for every $x \in E$,

$$\Big\|x - \sum_{i=1}^{n} \langle x, e_i\rangle e_i\Big\|^2 = \|x\|^2 - \sum_{i=1}^{n} \langle x, e_i\rangle^2. \tag{3.3.1}$$

We have already proved equality (3.3.1) in the course of Theorem 3.1.1: recall that $\sum_{i=1}^{n} \langle x, e_i\rangle e_i$ is precisely the **orthogonal projection** of x onto the subspace spanned by the vectors $e_1, \dots, e_n$. However, it can be useful to have a direct proof of (3.3.1), based on an immediate use of the properties of the inner product.

For any $c_1, \dots, c_n \in \mathbb{R}$ we have

$$\begin{aligned}\left\|x - \sum_{i=1}^{n} c_i e_i\right\|^2 &= \left\langle x - \sum_{i=1}^{n} c_i e_i, x - \sum_{i=1}^{n} c_i e_i\right\rangle \\ &= \langle x, x\rangle - \left\langle x, \sum_{i=1}^{n} c_i e_i\right\rangle - \left\langle \sum_{i=1}^{n} c_i e_i, x\right\rangle + \left\langle \sum_{i=1}^{n} c_i e_i, \sum_{j=1}^{n} c_j e_j\right\rangle\end{aligned}$$

$$= \|x\|^2 - 2\sum_{i=1}^{n} c_i\langle x, e_i\rangle + \sum_{i=1}^{n}\sum_{j=1}^{n} c_i c_j\langle e_i, e_j\rangle$$
$$= \|x\|^2 - 2\sum_{i=1}^{n} c_i\langle x, e_i\rangle + \sum_{i=1}^{n} c_i^2.$$

Taking $c_i = \langle x, e_i\rangle$ we obtain (3.3.1).

Let us now consider the properties of orthonormal infinite sequences, often called (countable) **orthonormal systems.**

2. Bessel's inequality and Parseval's identity

Proposition 3.3.1 (Bessel's inequality). *Let $(e_n)_{n\in\mathbb{N}}$ be an orthonormal sequence in an inner product space E and let $x \in E$. Then the series $\sum_{i=1}^{\infty}\langle x, e_i\rangle^2$ converges, and moreover we have*

$$\sum_{i=1}^{\infty}\langle x, e_i\rangle^2 \le \|x\|^2. \tag{3.3.2}$$

Proof. Equality (3.3.1) holds for each $n \in \mathbb{N}$ and shows that

$$\sum_{i=1}^{n}\langle x, e_i\rangle^2 \le \|x\|^2 \quad (n \in \mathbb{N}). \tag{3.3.3}$$

Letting $n \to \infty$, the statements in Proposition 3.3.1 follow immediately. □

Two problems

Question A

Given $x \in E$, does the series $\sum_{i=1}^{\infty}\langle x, e_i\rangle e_i$ of vectors of E converge?

Question B

Does the series above converge **to x**? In this case, that is, when the equality $\sum_{i=1}^{\infty}\langle x, e_i\rangle e_i = x$ holds, we speak of a **series expansion** of x into the orthonormal sequence $(e_n)_{n\in\mathbb{N}}$.

Proposition 3.3.2. *Let $(e_n)_{n\in\mathbb{N}}$ be an orthonormal sequence in an inner product space E and let $x \in E$. Then the following statements are* ***equivalent:***

$$\textbf{(a)} \quad \sum_{i=1}^{\infty}\langle x, e_i\rangle e_i = x, \tag{3.3.4}$$

$$\textbf{(b)} \quad \sum_{i=1}^{\infty}\langle x, e_i\rangle^2 = \|x\|^2. \tag{3.3.5}$$

Proof. This follows again from (3.3.1), just letting $n \to \infty$. □

Remark. The equality in (3.3.5) is called **Parseval's identity** and represents the special case in which Bessel's inequality becomes in fact an equality.

3. Total orthonormal systems

We recall (see Remark 3.2.11) that if (X, d) is any metric space and $A \subset X$, the set of all cluster points of A is denoted with $\overline{A}$ and called the *closure* of A.

Definition 3.3.1. A set $A \subset X$ is said to be **dense** in X if $\overline{A} = X$.

Thus, saying that A is dense in X means that every point of X is a cluster point of A, or in symbols

$$\forall x \in X \quad \forall \epsilon > 0 \quad \exists z \in A : d(z, x) < \epsilon.$$

For example, the set $\mathbb{Q}$ of the rational numbers is a dense subset of $\mathbb{R}$.

Theorem 3.3.1. *Let $(e_n)_{n\in\mathbb{N}}$ be an orthonormal sequence in an inner product space E. Suppose that $(e_n)_{n\in\mathbb{N}}$ is* total, *that is, the vector subspace spanned by $(e_n)_{n\in\mathbb{N}}$ is* dense *in E. Then*

$$\sum_{i=1}^{\infty} \langle x, e_i \rangle e_i = x \quad \forall x \in E. \tag{3.3.6}$$

Proof. The proof consists in using the totality of $(e_n)_{n\in\mathbb{N}}$ together with the best approximation property of the orthogonal projection.

Let $x \in E$. To prove (3.3.6), we need to show that

$$\forall \epsilon > 0 \quad \exists n_0 \in \mathbb{N} : \forall n \geq n_0 \quad \left\| x - \sum_{i=1}^{n} \langle x, e_i \rangle e_i \right\| < \epsilon. \tag{3.3.7}$$

Let M be the vector subspace of E spanned by $(e_n)_{n\in\mathbb{N}}$, that is,

$$M = \left\{ \sum_{i=1}^{n} c_i e_i \mid c_i \in \mathbb{R}, n \in \mathbb{N} \right\} = \bigcup_{n=1}^{\infty} M_n,$$

where

$$M_n = \operatorname{span}(e_1, \ldots, e_n).$$

Fix $\epsilon > 0$. Since M is dense in E by assumption, there exist an $\hat{n} \in \mathbb{N}$ and coefficients $c_i \in \mathbb{R}$ $(i = 1, \ldots, \hat{n})$ such that

$$\left\| x - \sum_{i=1}^{\hat{n}} c_i e_i \right\| < \epsilon. \tag{3.3.8}$$

However, by the nearest point property of the orthogonal projection, we have

$$\left\| x - \sum_{i=1}^{\hat{n}} \langle x, e_i \rangle e_i \right\| \le \left\| x - \sum_{i=1}^{\hat{n}} c_i e_i \right\|. \tag{3.3.9}$$

Moreover, by the same reason for every $n \ge \hat{n}$, since $M_{\hat{n}} \subset M_n$,

$$\left\| x - \sum_{i=1}^{n} \langle x, e_i \rangle e_i \right\| \le \left\| x - \sum_{i=1}^{\hat{n}} \langle x, e_i \rangle e_i \right\|. \tag{3.3.10}$$

Using (3.3.8), (3.3.9), and (3.3.10), the result follows. □

Remark 3.3.1. If $\sum_{i=1}^{\infty} \langle x, e_i \rangle e_i = x$ for every $x \in E$, then necessarily the vector subspace M spanned by $(e_n)_{n \in \mathbb{N}}$ is dense in E; indeed, consider that this equality, which corresponds to (3.3.7), gives much more detailed information than just the density of M. In conclusion, the validity of (3.3.6) is **equivalent** to the totality of $(e_n)_{n \in \mathbb{N}}$ (and, on the other side, to the validity of Parseval's identity (3.3.5) for every $x \in E$).

Remark 3.3.2. We have shown practically all the results above in the special case of the **trigonometric system**. In particular, among the results of the previous section (Fourier series), we have shown that the vector space spanned by the trigonometric functions, that is, the set of all the **trigonometric polynomials**, is dense in $\hat{C}_{2\pi}(\mathbb{R})$; in other words, the trigonometric system is **total**.

The sequence spaces l^p

Consider the set S of all real sequences

$$a = (a_1, a_2, \dots, a_n, \dots), \quad a_i \in \mathbb{R} \quad \forall i \in \mathbb{N}.$$

S has a natural structure of real vector space, for we can define the **sum** of two sequences and the **product** of a sequence by a scalar in the obvious way, that is, **componentwise**: if $a = (a_1, a_2, \dots, a_n, \dots)$ and $b = (b_1, b_2, \dots, b_n, \dots)$, then

$$a + b \equiv (a_1 + b_1, a_2 + b_2, \dots, a_n + b_n, \dots)$$

and

$$\lambda a \equiv (\lambda a_1, \lambda a_2, \dots, \lambda a_n, \dots).$$

Some vector subspaces of S are of interest. Given any real number p with $p \ge 1$, set

$$l^p = \left\{ a = (a_n) \in S : \sum_{n=1}^{\infty} |a_n|^p < +\infty \right\}.$$

It is customary to consider also the case $p = \infty$, which is defined in a different way:

$$l^\infty = \{a = (a_n) \in S : (a_n) \text{ is bounded}\}.$$

Exercise 3.3.1. Prove that for any $p : 1 < p < \infty$ we have

$$l^1 \subset l^p \subset l^\infty.$$

More generally, if $1 \le p, q \le \infty$ and $p < q$, then $l^p \subset l^q$.

Question. Is l^p a vector space?

The answer to this question rests on the following important inequality holding for non-negative sequences (a_n), (b_n) and for any $p \ge 1$:

$$\left(\sum_{n=1}^{\infty}(a_n + b_n)^p\right)^{\frac{1}{p}} \le \left(\sum_{n=1}^{\infty} a_n^p\right)^{\frac{1}{p}} + \left(\sum_{n=1}^{\infty} b_n^p\right)^{\frac{1}{p}}. \tag{3.3.11}$$

Inequality (3.3.11) is the special version for series of the famous *Minkowski inequality* (see equation (3.4.3)), holding in general L^p spaces. It follows that setting, for $a \in l^p$,

$$\|a\|_p \equiv \left(\sum_{n=1}^{\infty} |a_n|^p\right)^{\frac{1}{p}} \tag{3.3.12}$$

we get not only the fact that $a + b \in l^p$ if $a, b \in l^p$, but also that (3.3.12) defines a **norm** in the vector space l^p.

In fact, one proves the following result, which will also be stated in more general form in the next section.

Theorem 3.3.2. *For every p with $1 \le p < \infty$, equipped with the norm* (3.3.12), *l^p is a Banach space.*

Remark 3.3.3. Also l^∞ is a Banach space when equipped with the norm

$$\|a\|_\infty \equiv \sup_{n\in\mathbb{N}} |a_n|.$$

Indeed, this is nothing but the special case $A = \mathbb{N}$ of the Banach space $(B(A), \|\cdot\|)$, with

$$\|f\| = \sup_{x\in A} |f(x)| \quad \text{for } f \in B(A).$$

The special case $p = 2$

We have

$$l^2 = \left\{a = (a_n) \in S : \sum_{n=1}^{\infty} a_n^2 < +\infty\right\}.$$

Inner product in l^2

Proposition 3.3.3. *For every $a = (a_n), b = (b_n) \in l^2$ put*

$$\langle a, b\rangle \equiv \sum_{n=1}^{\infty} a_n b_n. \tag{3.3.13}$$

Then (3.3.13) *defines an inner product in l^2.*

Proof. This is left as an exercise. One has first to check that (3.3.13) is a good definition, namely, that the series in (3.3.13) is convergent. In fact, we can see that it is *absolutely* convergent, for using the inequality

$$|xy| \le \frac{1}{2}(x^2 + y^2) \quad (x, y \in \mathbb{R})$$

it follows that

$$\sum_{n=1}^{\infty} |a_n b_n| \le \frac{1}{2}\left(\sum_{n=1}^{\infty} a_n^2 + \sum_{n=1}^{\infty} b_n^2\right) < \infty. \qquad \square$$

Norm induced by the inner product

We have

$$\|a\| = \sqrt{\langle a, a\rangle} = \sqrt{\sum_{n=1}^{\infty} a_n^2} = \|a\|_2. \tag{3.3.14}$$

That this is a norm follows directly from the general properties of the inner product and does not require to be proved independently as for the case $p \neq 2$.

A prototype orthonormal system

For every $n \in \mathbb{N}$, put

$$e_n = (0, 0, \dots, 0, 1, 0, \dots),$$

where the number 1 appears only at the nth place of the sequence, that is, $(e_n)_j = 0$ if $j \neq n$ and $(e_n)_j = 1$ if $j = n$.

It is clear from (3.3.13) and (3.3.14) that

$$\begin{cases} \langle e_n, e_m\rangle = 0, & (n \neq m) \\ \langle e_n, e_n\rangle = \|e_n\|^2 = 1 \end{cases} \tag{3.3.15}$$

so that $(e_n)_{n\in\mathbb{N}}$ is an orthonormal system in l^2, which we can see as the l^2 version of the familiar orthonormal basis $(e_1, \dots, e_k)$ of $\mathbb{R}^k$.

Moreover, this orthonormal system is **total**: in fact, for every $a = (a_n) \in l^2$ we have

$$\langle a, e_n \rangle = a_n \quad \forall n \in \mathbb{N}$$

so that

$$\|a\|^2 = \sum_{n=1}^{\infty} a_n^2 = \sum_{n=1}^{\infty} \langle a, e_n \rangle^2. \tag{3.3.16}$$

This shows that Parseval's identity (3.3.5) holds for every $a \in l^2$ and thus proves the claim.

Exercise 3.3.2. Show that the orthogonal projection of $a = (a_n)_{n \in \mathbb{N}} \in l^2$ onto the subspace spanned by the first k vectors $e_1, \ldots, e_k$ of the sequence $(e_n)_{n \in \mathbb{N}}$ is

$$(a_1, a_2, \ldots, a_k, 0, 0 \ldots).$$

3.4 Hilbert spaces. Projection onto a closed convex set

Definition 3.4.1. A *Hilbert space* is an inner product space that is *complete* with respect to the distance induced by the inner product.

Example 3.4.1 (See Example 3.1.1). The basic example of Hilbert space is provided by the Euclidean space $\mathbb{R}^n$ equipped with the inner product (3.1.1).

Example 3.4.2 (See Example 3.1.2). The space $C([a, b])$, equipped with the inner product

$$\langle f, g \rangle = \int_a^b f(x) g(x)\, dx, \tag{3.4.1}$$

is *not* complete (the sequence $(\arctan nx)$ yields an example of a Cauchy sequence in $C([-1, 1])$ that does not converge). In order to complete it, it is necessary to enlarge the class of functions to be considered, first generalizing the definition of integral and then introducing the L^p spaces. To learn about these fundamental extensions, a classical reference is W. Rudin's book [6] (see also [10]); for the reader's convenience we only list some basic definitions and ideas in this context.

The Lebesgue integral and the L^p spaces

1. Measure and integration

A *measure space* is a triple (X, M, μ) where:
- X is a non-empty set;
- M is a *σ-algebra* of subsets of X, that is, a family of subsets of X containing X itself and closed with respect to the set operations of complement and countable union; the elements of M are called the *measurable subsets* of X;

– μ is a *measure* on M, that is, a function $\mu : M \to [0, +\infty]$ that is *countably additive*, i. e., such that

$$\mu\left(\bigcup_{n=1}^{\infty} E_n\right) = \sum_{n=1}^{\infty} \mu(E_n)$$

whenever (E_n) is a sequence of measurable sets with $E_n \cap E_m = \emptyset \; \forall n \neq m$.

A function $f : X \to \mathbb{R}$ is said to be *measurable* if the inverse image

$$f^{-1}(V) \equiv \{x \in X : f(x) \in V\}$$

is a measurable set for any open subset $V \subset \mathbb{R}$.

Given a measurable $f : X \to \mathbb{R}$ with $f \geq 0$ on X, it is possible to define the

$$\textbf{integral} \quad \int_E f \, d\mu$$

of f on the measurable set $E \subset X$ with respect to the measure μ: it is either a non-negative number (when, for instance, f is bounded and the measure $\mu(E)$ of E is finite) or $+\infty$; one then defines the integral for sign-changing functions. The integral so defined not only possesses all the usual properties of the Riemann integral (linearity, monotonicity with respect to the function, monotonicity with respect to the set, etc.) but is in fact much more general and flexible than the Riemann integral. Moreover, it reduces to it when considered for a continuous function on a compact interval of $\mathbb{R}$ and when one employs the *Lebesgue measure* of the (Lebesgue measurable) subsets of $\mathbb{R}$, constructed starting from the intervals and putting $m(I) = b - a$ if I is an interval of endpoints $a < b$.

2. The L^p spaces

Given any measure space (X, M, μ) and any real number p with $p \geq 1$, set

$$L^p(X, M, \mu) = \left\{f : X \to \mathbb{R} | f \text{ measurable}, \int_X |f|^p \, d\mu < \infty\right\}.$$

More generally, one can define $L^p(A, M_A, \mu_A)$, where A is a measurable subset of X (i. e., $A \in M$) and

$$M_A \equiv \{E \in M : E \subset A\}, \quad \mu_A \equiv \mu|_{M_A}.$$

We have two fundamental inequalities (called respectively *Hölder's inequality* and *Minkowski's inequality*) holding for any two non-negative measurable functions $f, g : X \to \mathbb{R}$ and for any $p > 1$:

$$\int_X fg\, d\mu \le \left(\int_X f^p\, d\mu\right)^{\frac{1}{p}} \left(\int_X g^q\, d\mu\right)^{\frac{1}{q}}, \tag{3.4.2}$$

where q is the *conjugate exponent* to p, in the sense that $\frac{1}{p} + \frac{1}{q} = 1$, and

$$\left(\int_X (f+g)^p\, d\mu\right)^{\frac{1}{p}} \le \left(\int_X f^p\, d\mu\right)^{\frac{1}{p}} + \left(\int_X g^p\, d\mu\right)^{\frac{1}{p}}. \tag{3.4.3}$$

The Minkowski inequality clearly holds also for $p = 1$ and shows first that, for every fixed $p \ge 1, f + g \in L^p$ if $f, g \in L^p$; moreover, it proves that setting, for $f \in L^p$,

$$\|f\|_p \equiv \left(\int_X |f|^p\, d\mu\right)^{\frac{1}{p}}, \tag{3.4.4}$$

we define a **norm** in the vector space L^p. Actually, to obtain this (and in particular the implication $\|f\|_p = 0 \Leftrightarrow f = 0$) one must identify functions that are equal *almost everywhere* on X; two measurable functions f, g are said to have this property if

$$\mu(\{x \in X : f(x) \neq g(x)\}) = 0.$$

In fact one proves the following.

Theorem 3.4.1. *For every measure space* (X, M, μ) *and for every* p *with* $1 \le p < \infty$, *the space* $L^p(X, M, \mu)$, *equipped with the norm* (3.4.4), *is a Banach space.*

Special cases

- Take $X = \mathbb{R}^n$ and

$$M = L \equiv \{E \subset \mathbb{R}^n : E \text{ is } \textit{Lebesgue measurable}\},$$
$$\mu = m : L \to [0, \infty] \equiv \text{the } \textit{Lebesgue measure}\,.$$

  Open sets (and hence also closed sets) in $\mathbb{R}^n$ are all Lebesgue measurable. It follows that any continuous function $f : \mathbb{R}^n \to \mathbb{R}$ is Lebesgue measurable.
- Take $X = \mathbb{N}$ and

$$M = M_0 \equiv \{E | E \subset \mathbb{N}\},$$
$$\mu = \nu \equiv \text{the counting measure},$$

  that is, $\nu(E)$ equals the number of elements of E if E is finite, while $\nu(E) = +\infty$ if E is infinite. One checks that, for this measure space and for any non-negative $a : \mathbb{N} \to \mathbb{R}$, that is, for any real sequence (a_n), one has

$$\int_{\mathbb{N}} a dv = \sum_{n=1}^{\infty} a_n$$

and it then follows that

$$L^p(\mathbb{N}, M_0, v) = l^p \quad \forall p \geq 1,$$

where the sequence spaces l^p have been defined in the previous section.

Remark 3.4.1. If Ω is a bounded open set in $\mathbb{R}^n$ and $C(\overline{\Omega})$ is defined as in (2.4.10), we have

$$C(\overline{\Omega}) \subset L^p(\Omega) \quad \text{for every } p \geq 1. \tag{3.4.5}$$

Indeed, if $f \in C(\overline{\Omega})$, f is measurable as already noted above, and moreover (as $|f(x)| \leq K$ in Ω for some $K \geq 0$) we also have

$$\int_{\Omega} |f|^p \, dm \leq \int_{\Omega} K^p \, dm = K^p m(\Omega) < \infty \tag{3.4.6}$$

because the Lebesgue measure of a (measurable) bounded set is finite. One fundamental property of the inclusion (3.4.5) is that for every $p \geq 1$, $C(\overline{\Omega})$ is a **dense** subset of $L^p(\Omega)$; this means (see Remark 3.2.11 and Definition 3.3.1) that every $f \in L^p(\Omega)$ can be approximated as closely as we wish (in the L^p norm!) with a function $g \in C(\overline{\Omega})$.

Remark 3.4.2. Consider again a bounded open set $\Omega \subset \mathbb{R}^n$. Using Hölder's inequality it is easy to see that

$$L^p(\Omega) \subset L^q(\Omega) \quad \text{if } q < p \tag{3.4.7}$$

and, in particular, that

$$\|f\|_q \leq C\|f\|_q \quad \forall f \in L^p(\Omega), \quad C = m(\Omega)^{\frac{1}{q}-\frac{1}{p}}. \tag{3.4.8}$$

Indeed, using (3.4.2) with conjugate exponents $\frac{p}{q}$ and $\frac{p}{p-q}$ we have

$$\int_{\Omega} |f|^q \, dm \leq \left(\int_{\Omega} |f|^p \, dm \right)^{\frac{q}{p}} \left(\int_{\Omega} 1 \, dm \right)^{\frac{p-q}{p}}$$
$$= \|f\|_p^q m(\Omega)^{\frac{p-q}{p}},$$

whence we obtain (3.4.8). It is clear that the same conclusions hold, more generally, whenever Ω is a measure space with measure $\mu(\Omega) < \infty$.

Exercise 3.4.1. Give examples showing that the inclusion (3.4.7) is false if the boundedness assumption on Ω is dropped.

3. The special case $p = 2$
We have

$$L^2(X, M, \mu) = \left\{f : X \to \mathbb{R} | f \text{ measurable}, \int_X f^2 \, d\mu < \infty\right\}.$$

Inner product in L^2: This is defined putting for every $f, g \in L^2$

$$\langle f, g\rangle \equiv \int_X fg \, d\mu. \tag{3.4.9}$$

Note that the **norm induced by the inner product**, that is,

$$\|f\| = \sqrt{\langle f, f\rangle} = \sqrt{\int_X f^2 \, d\mu}, \tag{3.4.10}$$

equals the norm $\|f\|_2$ defined in (3.4.4). Therefore, Theorem 3.4.1 yields the following corollary.

Corollary 3.4.1. *For every measure space (X, M, μ), the space $L^2(X, M, \mu)$, equipped with the inner product* (3.4.9), *is a Hilbert space.*

In particular, the sequence space l^2 considered in Section 3.3 is a Hilbert space.

Additional results for Hilbert spaces
In the next few pages we discuss some results holding in the context of Hilbert spaces and complementing those seen in the previous sections about the existence of the nearest point in a set K to a given point x, and in particular of the orthogonal projection of x onto a subspace M, and the properties of countable orthonormal systems $(e_n)_{n\in\mathbb{N}}$. Each of the two main results that we are going to prove, Theorem 3.4.2 and Theorem 3.4.3, will clearly display the role played by the **completeness** of the space.

Recall that a subset K of a vector space E is said to be **convex** if for any two points $x, y \in K$ we have

$$(1 - t)x + ty \in K \quad \forall t \in [0, 1].$$

Geometrically, this means that the whole **segment** of endpoints x, y is contained in K if $x, y \in K$.

Any vector subspace of E is convex. If E has a norm, then any open ball and any closed ball in E is convex.

Theorem 3.4.2 (Existence and uniqueness of the nearest point in a closed convex set). *Let H be a Hilbert space and let K be a* **closed** *and* **convex** *subset of H. Then for every $x \in H$, there exists a* unique *$z_0 \in K$ such that*

$$\|x - z_0\| \le \|x - z\| \quad \forall z \in K. \tag{3.4.11}$$

The theorem says that given any fixed $x \in H$, the function f of H into $\mathbb{R}$ defined putting

$$f(z) = \|x - z\|$$

attains its minimum on K at a unique point $z_0 \in K$. Thus, we need to concentrate on the non-negative number

$$\delta \equiv \inf_{z \in K} \|x - z\|, \tag{3.4.12}$$

usually called the **distance of** x **from** K, and see if this infimum is actually attained at some $z_0 \in K$, so as to satisfy (3.4.11). With the help of Exercise 3.1.3 (ensuring that f is continuous), the Weierstrass theorem immediately guarantees this provided that $H = \mathbb{R}^n$ and K is closed and bounded; in fact, Proposition 3.1.8 proves that closedness is enough in this case. How to proceed in the general case? Compactness would work; however, there are "few" compact sets in infinite dimensions. The completeness of H and the convexity of K will be the key ingredients for the proof of Theorem 3.4.2.

Lemma 3.4.1. *Let E be an inner product space and let K be a convex subset of E. Given $x \in E$, let δ be its distance from K defined in* (3.4.12). *Then we have*

$$\|z_1 - z_2\|^2 \le 2\|x - z_1\|^2 + 2\|x - z_2\|^2 - 4\delta^2 \quad \forall z_1, z_2 \in K. \tag{3.4.13}$$

Proof. Start from the parallelogram law

$$\|x + y\|^2 + \|x - y\|^2 = 2\|x\|^2 + 2\|y\|^2 \tag{3.4.14}$$

and rewrite it in the form

$$\|x - y\|^2 = 2\|x\|^2 + 2\|y\|^2 - 4\left\|\frac{x + y}{2}\right\|^2. \tag{3.4.15}$$

Writing this with $x - z_1$ in place of x and $x - z_2$ in place of y we obtain

$$\|z_1 - z_2\|^2 = 2\|x - z_1\|^2 + 2\|x - z_2\|^2 - 4\left\|x - \frac{z_1 + z_2}{2}\right\|^2, \tag{3.4.16}$$

whence we obtain (3.4.13) by virtue of the convexity of K. □

Proof of Theorem 3.4.2

Given $x \in H$, let again $\delta = \inf_{z \in K} \|x - z\|$ be its distance from K defined in (3.4.12).

1. (Existence) There exists a sequence $(z_n) \subset K$ such that

$$\|x - z_n\| \to \delta \quad \text{as} \quad n \to \infty \tag{3.4.17}$$

(see Exercise 3.1.4). We claim that (z_n) is a Cauchy sequence; indeed, (3.4.13) shows that

$$\|z_n - z_m\|^2 \le 2\|x - z_n\|^2 + 2\|x - z_m\|^2 - 4\delta^2 \quad \forall n, m \in \mathbb{N}. \tag{3.4.18}$$

Using (3.4.17) we then see that

$$\|z_n - z_m\| \to 0 \quad \text{as} \quad n, m \to \infty,$$

so that our claim is proved. Thus, by the completeness of H, there is a $z_0 \in H$ such that

$$z_n \to z_0 \quad (n \to \infty).$$

However, K is a closed set by assumption, so $z_0 \in K$ (Exercise 3.1.2).
Moreover, by continuity of the norm we have

$$\|x - z_n\| \to \|x - z_0\| \quad (n \to \infty)$$

so that using again (3.4.17), we finally obtain $\|x - z_0\| = \delta$, whence the result follows.

2. (Uniqueness) Suppose that there exist $z_1, z_2 \in K$ such that

$$\|x - z_1\| = \|x - z_2\| = \delta.$$

Then using (3.4.13) we see immediately that $z_1 = z_2$.

Corollary 3.4.2 (Orthogonal projection onto a closed subspace). *Let H be a Hilbert space and let M be a* ***closed*** *vector subspace of H. Then for every $x \in H$, there exists a unique $y \in M$ such that*

$$\|x - y\| \le \|x - z\| \quad \forall z \in M. \tag{3.4.19}$$

Moreover, y is ***characterized*** *by the property that*

$$x - y \perp M. \tag{3.4.20}$$

Proof. The first statement is just the special case of Theorem 3.4.2 where K is a vector subspace of H. The second statement is a mere repetition of the equivalence between (3.4.19) and (3.4.20) already proved in Section 3.1; see in particular Remark 3.1.4 and Proposition 3.1.7. □

Orthonormal systems in a Hilbert space

Let $(e_n)_{n\in\mathbb{N}}$ be an orthonormal system in a Hilbert space. What can be added to the results seen in the previous section in the context of inner product spaces? The next theorem shows the role played by completeness in the convergence of the series of interest.

Theorem 3.4.3 (Fischer–Riesz theorem). *Let $(e_n)_{n\in\mathbb{N}}$ be an orthonormal sequence in a Hilbert space H. Then for every $x \in H$, the series $\sum_{n=1}^{\infty}\langle x, e_n\rangle e_n$ converges in H, and moreover putting*

$$p(x) \equiv \sum_{n=1}^{\infty}\langle x, e_n\rangle e_n \tag{3.4.21}$$

we have

$$\langle p(x), e_i\rangle = \langle x, e_i\rangle \quad \forall i \in \mathbb{N}. \tag{3.4.22}$$

Proof.

- We first prove the following statement: for every $a = (a_n) \in l^2$, the series $\sum_{n=1}^{\infty} a_n e_n$ converges in H, and if $p = \sum_{n=1}^{\infty} a_n e_n$, then

$$\langle p, e_i\rangle = a_i \quad \forall i \in \mathbb{N}. \tag{3.4.23}$$

  To do this, set

$$s_n = \sum_{i=1}^{n} a_i e_i, \quad t_n = \sum_{i=1}^{n} a_i^2 \quad (n \in \mathbb{N}).$$

  We have, assuming for instance $n > m$,

$$\begin{aligned}\|s_n - s_m\|^2 &= \left\|\sum_{i=1}^{n} a_i e_i - \sum_{i=1}^{m} a_i e_i\right\|^2 = \left\|\sum_{i=m+1}^{n} a_i e_i\right\|^2 \\ &= \sum_{i=m+1}^{n}\|a_i e_i\|^2 = \sum_{i=m+1}^{n}|a_i|^2 = |t_n - t_m|.\end{aligned}$$

  By the assumption on a, this shows that (s_n) is a Cauchy sequence in H and therefore converges. Equality (3.4.23) is a consequence of the continuity of the scalar product; see Exercise 3.4.2 below.
- Now let $x \in H$. By Bessel's inequality (Proposition 3.3.1), we know that the sequence $(\langle x, e_n\rangle)_{n\in\mathbb{N}} \in l^2$; therefore, the statements in the theorem follow as a special case from the above discussion. □

Exercise 3.4.2. Let E be an inner product space.

- The inequality

$$\left|\langle x - y, a\rangle\right| \le \|x - y\|\|a\|$$

  shows that the (linear) map $x \to \langle x, a\rangle$ ($a \in E$ fixed) of E into $\mathbb{R}$ is continuous.

– It follows that for any convergent series $\sum_{n=1}^{\infty} x_n$ in E and for every fixed $a \in E$, we have

$$\left\langle \sum_{n=1}^{\infty} x_n, a \right\rangle = \sum_{n=1}^{\infty} \langle x_n, a \rangle.$$

Exercise 3.4.3. The closure $\overline{M}$ of a vector subspace M of a normed vector space is a closed vector subspace (use Remark 3.2.11).

Corollary 3.4.3. *Let $(e_n)_{n\in\mathbb{N}}$ be an orthonormal sequence in a Hilbert space H and let $x \in H$. Then the vector $p(x)$ defined by formula* (3.4.21) *is the orthogonal projection of x onto the closure $\overline{M}$ of the vector subspace M spanned by $(e_n)_{n\in\mathbb{N}}$.*

Proof. Given $x \in H$, the existence and uniqueness of its orthogonal projection onto $\overline{M}$ is established by Corollary 3.4.2. To prove our assertion, it is therefore enough to show that

$$x - p(x) \perp \overline{M}. \tag{3.4.24}$$

In fact, using (3.4.22) we have

$$\langle x - p(x), e_i \rangle = \langle x, e_i \rangle - \langle p(x), e_i \rangle = \langle x, e_i \rangle - \langle x, e_i \rangle = 0$$

for every $i \in \mathbb{N}$, which shows that $x - p(x) \perp M$. Formula (3.4.24) follows on using once again the continuity of the inner product (Exercise 3.4.2).

As a special case we obtain the following statement, which is a weaker form of Theorem 3.3.1 on the series expansion of every vector of the space. □

Corollary 3.4.4. *Let $(e_n)_{n\in\mathbb{N}}$ be an orthonormal sequence in a Hilbert space H. Suppose that $(e_n)_{n\in\mathbb{N}}$* is total, *that is, the vector subspace spanned by $(e_n)_{n\in\mathbb{N}}$* is dense *in E. Then for every $x \in H$ we have*

$$\sum_{i=1}^{\infty} \langle x, e_i \rangle e_i = x. \tag{3.4.25}$$

Proof. The assumption on $(e_n)_{n\in\mathbb{N}}$ means that $\overline{M} = H$, and it is clear that the orthogonal projection $p(x)$ of x onto H is x itself. □

Remark 3.4.3. In the light of the results discussed in this section, it is remarkable that Theorem 3.3.1 holds *independently* of the completeness of the space.

Exercise 3.4.4. Let $(e_n)_{n\in\mathbb{N}}$ be an orthonormal sequence in a Hilbert space H. Define a map F of H into the space S of all real sequences as follows:

$$F(x) \equiv (\langle x, e_n \rangle)_{n\in N}.$$

Show the following:

- F is a map of H **into** l^2;
- $F : H \to l^2$ is linear and continuous;
- F is surjective.

Moreover, if $(e_n)_{n\in\mathbb{N}}$ is total (in which case it is sometimes called a **Hilbert basis** of H), then F is also injective, and thus an *isomorphism* of H onto l^2. This shows that every Hilbert space with a (countable) Hilbert basis is **isomorphic to** l^2, which also shows the peculiar importance of this sequence space.

3.5 Additions and exercises

A1. Some remarks on orthonormal systems

1. In Section 3.2 of the present chapter we have seen by direct computation that the (full) trigonometric system

$$e_0 = \frac{1}{\sqrt{2\pi}}, \quad e_{2k-1}(x) = \frac{\cos kx}{\sqrt{\pi}}, \quad e_{2k}(x) = \frac{\sin kx}{\sqrt{\pi}} \quad (k \in \mathbb{N}) \tag{3.5.1}$$

is orthonormal in $C([0, 2\pi])$ (or $C([-\pi, \pi])$) equipped with the scalar product

$$\langle f, g\rangle = \int_0^{2\pi} f(x)g(x)\, dx, \tag{3.5.2}$$

and we have also seen (essentially, if not in all details) that this system is total (or **complete**, as is often said); these two properties are at the basis of the Fourier expansion of 2π-periodic functions.

2. The (only sine, partial) trigonometric system

$$\sqrt{\frac{2}{\pi}} \sin nx, \quad n \in \mathbb{N}, \tag{3.5.3}$$

is a total orthonormal system in $C([0, \pi])$ – if we take of course as scalar product the integral in (3.5.2) extended from 0 to π. This can be checked by methods entirely similar to those recalled in **1.**, that is, first direct computation, checking that

$$\int_0^{\pi} \sin^2 nx\, dx = \frac{\pi}{2} \quad \forall n \in \mathbb{N}, \quad \int_0^{\pi} \sin nx \sin mx\, dx = 0 \quad (n \neq m),$$

and then via the use of approximation methods based on the Weierstrass theorem. A slightly different method to verify the desired properties of the system (3.5.3) would consist in considering the **odd reflection** f_d of a function $f \in C([0, \pi])$ to $[-\pi, \pi]$ so as to return to **1.** and then reach the conclusions using the fact that the cosine coefficients

a_k of f_d are all 0; this is precisely the argument that we shall follow in Section 4.3 of Chapter 4.

In any case we obtain, using the orthonormality and totality of the system (3.5.3), the expansion

$$f = \sum_{n=1}^{\infty} \langle f, e_n \rangle e_n, \quad e_n(x) = \sqrt{\frac{2}{\pi}} \sin nx, \quad n \in \mathbb{N}, \tag{3.5.4}$$

which we can explicitly write as

$$f(x) = \sum_{n=1}^{\infty} \left(\int_0^{\pi} f(x) \sqrt{\frac{2}{\pi}} \sin nx \, dx \right) \cdot \sqrt{\frac{2}{\pi}} \sin nx,$$

that is, as

$$f(x) = \sum_{n=1}^{\infty} b_n \sin nx, \quad \text{with } b_n = \frac{2}{\pi} \int_0^{\pi} f(x) \sin nx \, dx \quad (n \in \mathbb{N}). \tag{3.5.5}$$

It should be clear to us that equality (3.5.5) **cannot** be interpreted literally as saying "for every $x \in [0, \pi]$, the series on the right-hand side of equation (3.5.5) converges to the sum $f(x)$" – that is, as asserting the **pointwise convergence** of the series to the given f; indeed, the general formula in (3.5.4) establishes the convergence of the series **in the norm of the given inner product space**, and thus in our case, the convergence is in the **quadratic mean** – not implying necessarily the pointwise convergence.

The pointwise, and in fact **uniform**, convergence of the series in (3.5.5) can be established if we assume further conditions on f besides the mere continuity; for instance, it will be shown in Proposition 4.4.2 that this is true if $f \in C_0^1[0, \pi]$.

A quite different method to establish the properties of the system (3.5.3) is based on the following fundamental remark. As we are going to see in the next exercise, the functions $\{\sin nx : n \in \mathbb{N}\}$ are the (non-normalized) **eigenfunctions** of the *boundary value problem* (BVP)

$$\begin{cases} u'' + \lambda u = 0 & (0 < x < \pi) \\ u(0) = u(\pi) = 0. \end{cases}$$

Exercise 3.5.1. Consider the BVP

$$(\mathbf{A}) \begin{cases} u'' + \lambda u = 0 & (a < x < b) \\ u(a) = u(b) = 0. \end{cases}$$

– Prove that **(A)** has solutions $u \neq 0$ iff

$$\lambda = \lambda_k = \left(\frac{\pi k}{b - a} \right)^2 \quad (k = 1, 2, \dots). \tag{3.5.6}$$

– Prove that the solutions corresponding to λ_k are given by

$$u_k(x) = C \sin \omega_k (x - a), \quad C \in \mathbb{R} \quad (\omega_k = \sqrt{\lambda_k}). \tag{3.5.7}$$

(i) The characteristic equation of the differential equation is $\mu^2 + \lambda = 0$. For $\lambda = 0$, **(A)** clearly only has the solution $u = 0$. Thus, consider the different cases:

(a) $\boldsymbol{\lambda < 0}$: Then $\mu^2 = -\lambda > 0$, so $\mu = \pm\sqrt{-\lambda} \equiv \pm\alpha$, $\alpha = \sqrt{-\lambda} > 0$.

The solutions of the differential equation are

$$u(x) = Ce^{\alpha x} + De^{-\alpha x}.$$

Imposing the boundary conditions

$$\begin{cases} u(a) = Ce^{\alpha a} + De^{-\alpha a} = 0 \\ u(b) = Ce^{\alpha b} + De^{-\alpha b} = 0, \end{cases}$$

we see that the determinant of the coefficient matrix is $e^{\alpha(a-b)} - e^{-\alpha(a-b)} \neq 0$, so $C = D = 0$.

(b) $\boldsymbol{\lambda > 0}$: Then $\mu^2 = -\lambda < 0$, so $\mu = \pm i\sqrt{\lambda} \equiv \pm i\omega$, $\omega = \sqrt{\lambda} > 0$.

The solutions of the differential equation are

$$u(x) = C \cos \omega x + D \sin \omega x. \tag{3.5.8}$$

Impose the boundary conditions

$$\begin{cases} u(a) = C \cos \omega a + D \sin \omega a = 0 \\ u(b) = C \cos \omega b + D \sin \omega b = 0. \end{cases} \tag{3.5.9}$$

The determinant of the coefficient matrix is

$$\cos \omega a \sin \omega b - \sin \omega a \cos \omega b = \sin \omega (b - a)$$

so that it equals zero iff

$$\sin \omega(b - a) = 0 \Leftrightarrow \omega(b - a) = k\pi \quad (k \in \mathbb{N}),$$

whence we finally get

$$\omega = \frac{k\pi}{b - a} \equiv \omega_k \quad (k \in \mathbb{N}).$$

As $\omega = \sqrt{\lambda}$, this gives (3.5.6).

(ii) From the first equation in (3.5.9) we obtain (assuming for instance $\cos\omega a \neq 0$) $C = -D\frac{\sin\omega a}{\cos\omega a}$, and putting this in (3.5.8) we have

$$\begin{aligned} u(x) &= -D\frac{\sin\omega a}{\cos\omega a}\cos\omega x + D\sin\omega x \\ &= D\left(\frac{-\sin\omega a\cos\omega x + \sin\omega x\cos\omega a}{\cos\omega a}\right) \\ &= \frac{D}{\cos\omega a}\sin\omega(x-a) \equiv D'\sin\omega(x-a). \end{aligned} \tag{3.5.10}$$

For $\omega = \omega_k$, we thus obtain (3.5.7).

The connection between total orthonormal systems and BVPs for second-order linear ODEs shown by Exercise 3.5.1 is far from being accidental: it is rather at the basis of what is called the **spectral theory** of (second-order) ordinary differential operators. We shall say something on this in the next Addition **A2**, while here below we just cite some keywords often encountered in the study of orthonormal systems.

3. Gram–Schmidt orthogonalization method: This is a standard constructive method – based on the properties of the orthogonal projection onto a subspace discussed in Section 3.1 of this chapter – allowing to pass from a set of linearly independent vectors $\{x_1,\dots,x_n\}$ in a inner product space E to a set of orthonormal vectors $\{u_1,\dots,u_n\}$ spanning the same subspace as the x_i; see, for instance, Chapter 3 of Kolmogorov–Fomin [9] or Chapter 3 of Taylor [8].

4. Special functions: It is fair to mention here some important orthonormal systems, useful in approximation theory and numerical analysis, some of which originate as eigenfunctions of (regular or singular) second-order linear ODEs; see, for instance, Chapter 7 of Kolmogorov–Fomin [9] or Chapter 7 of Weinberger [2].

(i) *Legendre polynomials*: These are obtained via orthogonalization of the functions $1, x, x^2, \dots$ on the interval $[-1, 1]$.
(ii) *Hermite polynomials*: These are obtained via orthogonalization on the interval $]-\infty,+\infty[$ of the functions $x^n e^{-\frac{x^2}{2}}$ (that is to say, the same functions considered in (i) multiplied by the **weight** $e^{-\frac{x^2}{2}}$, which allows these functions to be square integrable on $\mathbb{R}$).
(iii) *Laguerre polynomials*: Like in (ii), these are obtained once more by the integer power functions x^n, each multiplied by the weight e^{-x} and orthogonalized on $[0,+\infty[$.

A2. Some spectral theory for ordinary differential operators

We resume the notation and context of Addition A1 to Chapter 1, save that here – also in preparation for the content of Chapter 4 – we will denote the independent variable with x (rather than with t) and the unknown function in the ODE with u (rather than with x). Thus, put

$$Lu = u'' + a_1(x)u' + a_2(x)u, \tag{3.5.11}$$

where $a_1, a_2 \in C[a,b] \equiv C([a,b])$; in this addition, we use the simplified symbol for our convenience. Consider the homogeneous problem, depending on the (real) parameter λ,

$$\begin{cases} Lu - \lambda u = 0 & a < x < b \\ u(a) = 0 \\ u(b) = 0. \end{cases} \tag{3.5.12}$$

For every $\lambda \in \mathbb{R}$, (3.5.12) possesses of course the trivial solution $u = 0$. Values of λ for which (3.5.12) possesses also non-trivial solutions are called **eigenvalues** of the problem (3.5.12) (or *eigenvalues of L with Dirichlet boundary conditions* [BCs] *on u*), and the corresponding non-trivial solutions are called **eigenfunctions**.

Different boundary conditions can be considered; for instance, one can require that $u'(a) = u'(b) = 0$ (*Neumann* BCs) or that $u(a) = u'(b) = 0$, and so on. It is convenient to consider in general linear conditions involving u and u' at each endpoint of $[a,b]$ (**separated boundary conditions**) and thus consider the following more general form of (3.5.12):

$$\begin{cases} Lu - \lambda u = 0 & a < x < b \\ \alpha u(a) + \beta u'(a) = 0 \\ \gamma u(b) + \delta u'(b) = 0, \end{cases} \tag{3.5.13}$$

where

$$(\alpha, \beta) \neq (0,0) \quad \text{and} \quad (\gamma, \delta) \neq (0,0). \tag{3.5.14}$$

From the discussion in Addition A1 to Chapter 1, we gain immediately the following result concerning the non-homogeneous problem attached to (3.5.13).

Proposition 3.5.1. *Suppose that λ is **not** an eigenvalue of* (3.5.13). *Then the non-homogeneous problem*

$$\begin{cases} Lu - \lambda u = y(x) & a < x < b \\ \alpha u(a) + \beta u'(a) = m \\ \gamma u(b) + \delta u'(b) = n \end{cases} \tag{3.5.15}$$

has a solution for any given $y \in C[a,b]$ and any given $m, n \in \mathbb{R}$.

The definitions given above and the statement in Proposition 3.5.1 immediately open various questions, which we can roughly simplify as follows:
- Do there exist eigenvalues of the problem (3.5.13)? If so, "how many"?
- Which properties do the eigenfunctions attached to a given eigenvalue possess, and what can we say about the totality of them?

– What can we say about the solvability of problem (3.5.15) when λ is an eigenvalue of (3.5.13)?

The study of these questions, and of similar ones for much more general situations regarding linear operators in normed spaces, goes under the name of **spectral theory** [7, 11]. About (3.5.15), two simple points can be established right now, which are plain consequences of some linear algebra and what we have seen in Chapter 1.

(a) The set of all eigenfunctions corresponding to a given eigenvalue λ, together with the zero function, forms a vector space that is called the **eigenspace** corresponding to the eigenvalue λ. In our case, an eigenspace is a vector subspace of $C^2[a,b]$:

$$E(\lambda) = \{u \in C^2[a,b] | u \text{ is a solution of problem (3.5.13)}\},$$

and on the basis of the "dimensional theorem" (Theorem 1.3.1) we have

$$1 \le \dim E(\lambda) \le 2.$$

However, it is easily seen that $\dim E(\lambda) = 1$, for $\dim E(\lambda) = 2$ would mean that **every** solution of the linear second-order ODE $Lu = \lambda u$ has to satisfy the boundary conditions in (3.5.13), which is not true because the existence and uniqueness theorem for the IVP guarantees that we can choose the initial conditions without any constraint – indeed for instance, the solution z of the equation such that $z(a) = \alpha, z'(a) = \beta$ does not satisfy the BC at a.

(b) If λ is an eigenvalue of (3.5.13) and the problem (3.5.15) has a solution for a given y, then we are far away from uniqueness of this solution, for there are infinitely many of them; precisely, if we call $\overline{x}$ such a solution, then all functions in the "affine subspace"

$$S_y \equiv \overline{x} + E(\lambda)$$

have the same property. Of course, this is true for any equation of the type $A(x) - \lambda x = y$, with A a linear operator acting in a vector space E.

Much more can be said about the **Sturm–Liouville eigenvalue problems.**

These are of the form (3.5.13), with the following special type of differential operator L:

$$Lu = -(p(x)u')' + q(x)u, \tag{3.5.16}$$

where the coefficient functions satisfy the conditions

$$p \in C^1[a,b], \quad q \in C[a,b], \quad p(x) > 0 \quad \forall x \in [a,b].$$

It is useful to put

$$a(u,v) = \int_a^b p(x)u'(x)v'(x)\,dx + \int_a^b q(x)u(x)v(x)dx \quad (u,v \in C^1[a,b]); \tag{3.5.17}$$

a is often called the **Dirichlet form** corresponding to the differential operator L. Note that

$$a(u,v) = a(v,u) \quad \forall u,v$$

so that a is a **symmetric** bilinear form on the vector space $C^1[a,b]$; note in particular that – unlike L, which is naturally defined on $C^2[a,b]$ – a contains only first-order derivatives of the functions involved and is therefore well defined on the larger space $C^1[a,b]$. The relation between L and a – and the name of the latter – is made clear by the following.

Proposition 3.5.2. *We have*

$$\langle Lu, v\rangle = a(u,v) \quad \textit{for } u,v \in C_0^2[a,b], \tag{3.5.18}$$

where

$$C_0^2[a,b] \equiv \{u \in C^2[a,b] : u(a) = u(b) = 0\}.$$

Proof. Omitting for simplicity the x-dependence in the functions to be integrated, we can write

$$\begin{aligned} \langle Lu, v\rangle &= \int_a^b [-(pu')' + qu]v\,dx = -\int_a^b (pu')'v + \int_a^b quv\,dx \\ &= -pu'v]_{x=a}^{x=b} + \int_a^b pu'v'\,dx + \int_a^b quv\,dx = a(u,v), \end{aligned} \tag{3.5.19}$$

because the boundary term

$$B(u,v) \equiv pu'v]_{x=a}^{x=b} = p(b)u'(b)v(b) - p(a)u'(a)v(a)$$

vanishes, since $v(b) = v(a) = 0$. □

Note that only the boundary condition on v was used to obtain (3.5.18). Likewise, we see at once that $B(u,v) = 0$ for all v if u satisfies the (zero) Neumann BCs:

$$u'(a) = u'(b) = 0,$$

producing again equality (3.5.18). However, we look now for symmetric conditions on u and v ensuring the important property (3.5.18). Here the Sturm–Liouville separated BCs come into play, in view of the following remark.

Remark 3.5.1. If $u, v \in C^1[a,b]$ both satisfy a condition of the type

$$\alpha u(x_0) + \beta u'(x_0) = 0 \tag{3.5.20}$$

for some $x_0 \in [a,b]$ and some $(\alpha, \beta) \neq (0,0)$, then

$$u'(x_0)v(x_0) = u(x_0)v'(x_0). \tag{3.5.21}$$

One way to verify the statement above is just considering the linear algebraic system consisting of (3.5.20) and the same equation written for v, and concluding that the coefficient matrix must have determinant equal to zero.

Let us go back to the relation between L and a. Equation (3.5.19) shows that in general we have

$$\langle Lu, v\rangle = -B(u,v) + a(u,v). \tag{3.5.22}$$

Suppose now that the functions u, v satisfy both boundary conditions in (3.5.13). Then using the necessary condition (3.5.21) both at b and at a we get

$$\begin{aligned} B(u,v) &= p(b)u'(b)v(b) - p(a)u'(a)v(a) \\ &= p(b)u(b)v'(b) - p(a)u(a)v'(a) = B(v,u) \end{aligned}$$

and see that also B is symmetric in this case. The conclusion can then be stated as follows.

Proposition 3.5.3. *Let F be the vector subspace of $C^2[a,b]$ consisting of those functions satisfying both boundary conditions in* (3.5.13). *Then*

$$\langle Lu, v\rangle = \langle Lv, u\rangle \tag{3.5.23}$$

for every $u, v \in F$; in words, the bilinear form $(u,v) \to \langle Lu, v\rangle$ is ***symmetric*** *on F.*

The symmetry property (3.5.23) is crucial in dealing with the Sturm–Liouville eigenvalue problem (3.5.13) and also with the associated non-homogeneous problem (3.5.15). About the latter, we have the following statement.

Proposition 3.5.4. *Suppose that λ is an eigenvalue of* (3.5.13). *Then the non-homogeneous problem*

$$\begin{cases} Lu - \lambda u = y & a < x < b \\ \alpha u(a) + \beta u'(a) = 0 \\ \gamma u(b) + \delta u'(b) = 0 \end{cases} \tag{3.5.24}$$

can have a solution only if y is orthogonal to the solutions of the homogeneous problem (3.5.13), *that is, to the eigenfunctions corresponding to the eigenvalue λ.*

Proof. Suppose that u is a solution of (3.5.24) and let z be a solution of (3.5.13). Then

$$\begin{aligned}\langle y, z\rangle &= \langle Lu - \lambda u, z\rangle = \langle Lu, z\rangle - \lambda\langle u, z\rangle \\ &= \langle Lz, u\rangle - \lambda\langle z, u\rangle = \langle Lz - \lambda z, u\rangle = 0.\end{aligned}$$ □

Exercise 3.5.2. Interpret Example 1.7.3 of Chapter 1 in the light of Proposition 3.5.4.

Remark 3.5.2. It can be shown that the orthogonality condition stated in Proposition 3.5.4 is also **sufficient** for the solvability of (3.5.24).

The next result about (3.5.13) reminds us very neatly of an important property of the eigenvectors of real symmetric $n \times n$ matrices.

Proposition 3.5.5. *Eigenfunctions corresponding to different eigenvalues of* (3.5.13) *are orthogonal.*

Proof. We wish to prove that if

$$(*)\begin{cases} Lu = \lambda u \\ \alpha u(a) + \beta u'(a) = 0 \\ \gamma u(b) + \delta u'(b) = 0 \end{cases} \quad \text{and} \quad (**)\begin{cases} Lv = \mu v \\ \alpha v(a) + \beta v'(a) = 0 \\ \gamma v(b) + \delta v'(b) = 0 \end{cases}$$

with $\lambda \neq \mu$, then

$$\langle u, v\rangle = \int_a^b u(x)v(x)\,dx = 0.$$

Indeed, taking the scalar product by v of both members in the differential equation in (*) we have

$$\langle Lu, v\rangle = \lambda\langle u, v\rangle \tag{3.5.25}$$

and doing similar work with (**) we have

$$\langle Lv, u\rangle = \mu\langle v, u\rangle. \tag{3.5.26}$$

However, by virtue of equality (3.5.23) proved in Proposition 3.5.3, the first members in (3.5.25) and (3.5.26) are equal, implying that

$$(\lambda - \mu)\langle u, v\rangle = 0,$$

whence the desired conclusion follows.

On the basis of Proposition 3.5.5, we can thus say that *the (normalized) eigenfunctions of the Sturm–Liouville problem* (3.5.13) *form an orthonormal system in* $C[a, b]$; but this is evidently a vacuous statement until we are able to prove that eigenvalues to (3.5.13) actually exist! □

Existence of the eigenvalues

To answer this question, in the first instance one tries to extend what has been done in the basic Exercise 3.5.1. Consider for instance the more general eigenvalue problem (3.5.12) and let, for each $\lambda \in \mathbb{R}$, u_λ and v_λ be two independent solutions of the equation $Lu = \lambda u$, so that the general solution of this equation is

$$u = cu_\lambda + dv_\lambda \quad (c, d \in \mathbb{R}).$$

Impose the boundary conditions to determine c and d; for the Dirichlet BCs assigned in (3.5.12) we will write

$$\begin{cases} cu_\lambda(a) + dv_\lambda(a) = 0 \\ cu_\lambda(b) + dv_\lambda(b) = 0. \end{cases} \tag{3.5.27}$$

The linear algebraic system (3.5.27) can either have only the trivial solution $c = d = 0$ (meaning that λ is not an eigenvalue) or else have solutions $(c, d) \neq (0, 0)$; this happens precisely when

$$D(\lambda) = \det \begin{pmatrix} u_\lambda(a) & v_\lambda(a) \\ u_\lambda(b) & v_\lambda(b) \end{pmatrix} = 0. \tag{3.5.28}$$

It is thus a matter of studying the zeros of the function D. For instance, in Exercise 3.5.1 we had

$$u_\lambda(x) = \cos\sqrt{\lambda}x, \quad v_\lambda(x) = \sin\sqrt{\lambda}x$$

so that

$$D(\lambda) = \det \begin{pmatrix} \cos\sqrt{\lambda}\,a & \sin\sqrt{\lambda}\,a \\ \cos\sqrt{\lambda}\,b & \sin\sqrt{\lambda}\,b \end{pmatrix} = \sin\sqrt{\lambda}(b-a)$$

producing through its zero the eigenvalues $\lambda_k = (\frac{\pi k}{b-a})^2$, $k \in \mathbb{N}$.

Does a result like this hold for the general Sturm–Liouville problems (3.5.12) or (3.5.13)? The answer to this question is positive, but unfortunately this cannot be proved by the elementary methods used in this book. We can only indicate below some ways leading to the result.

One method is based on the theory of analytic functions of a complex variable. Widening the study of the ODE $x' = f(t, x)$ to the complex field $\mathbb{C}$, that is, studying the equation $z' = f(t, z)$, where now f is a function defined in a subset $A \subset \mathbb{R} \times \mathbb{C}$ and with values in $\mathbb{C}$, one may prove existence results for the solutions similar to those indicated in this book; references for this are, for instance, Coddington–Levinson [12] and

Dieudonné [7]. In particular, if f is *analytic* as a function of z, then the solution will depend analytically on the initial value z_0 ([12], Theorem 8.4) and on complex parameters λ appearing analytically in the equation. Applying this enlarged theory to equations such as that in (3.5.13), where now λ runs into the complex numbers, one proves that the solutions u_λ, v_λ depend analytically on λ for fixed t and are in fact entire functions of λ. It follows that the same is true for the determinant function $D(\lambda)$ defined in (3.5.28). In the words of [12], "This function can have only real zeros because (3.5.13) has no non-real eigenvalues. Thus, D is an entire function of λ which is not identically zero. Its zeros, which are the eigenvalues of (3.5.13), can therefore cluster only at $+\infty$."

Proofs of the statements above would clearly require a much deeper study of the matter, but we are not far from the truth saying that this method – due to the nature of the analytic functions as sum of convergent power series – is in principle the same that we use to show that a real symmetric matrix A has n real eigenvalues (counting multiplicities) appealing to the fact that the determinant

$$d(\lambda) \equiv \det(A - \lambda I)$$

is a polynomial of degree n in λ and so, by the fundamental theorem of algebra, it will have n zeros belonging to the complex field, which however will all lie in $\mathbb{R}$ due to the symmetry of A.

In the middle course of this quite informal discussion, it is now good and satisfying to state a standard form of the main results about (3.5.13), which gives a rather complete answer to our questions.

The spectral theorem for Sturm–Liouville eigenvalue problems

Theorem 3.5.1. *The Sturm–Liouville eigenvalue problem* (3.5.13) *has an infinite sequence* (λ_k) *of eigenvalues, which can be arranged in increasing order of magnitude,*

$$\lambda_1 < \lambda_2 < \cdots < \lambda_k < \cdots,$$

and are such that $\lambda_k \to +\infty$ *as* $k \to \infty$. *Each eigenvalue* λ_k *has multiplicity 1 (that is, eigenfunctions corresponding to* λ_k *are multiples of each other), and if* u_k *denotes a normalized eigenfunction corresponding to* λ_k, *then the* (u_k) *form a total orthonormal system in the space* $C([a,b])$ *equipped with the scalar product*

$$\langle u, v\rangle = \int_a^b u(x)v(x)\,dx.$$

A detailed proof of Theorem 3.5.1 can be found, for instance, in Dieudonné [7]. A more informal and computational approach is followed by H. Weinberger [2], which we now briefly discuss. First put

$$E_0 = \{u \in C([a,b]) \mid u \text{ is piecewise } C^1 \text{ and } u(a) = u(b) = 0\}, \tag{3.5.29}$$

where piecewise C^1 functions are defined in Definition 4.3.1. Then E_0 is a vector subspace of $C([a,b])$ and is equipped with the scalar product $\langle u, v\rangle = \int_a^b u(x)v(x)\,dx$ inherited by that of $C([a,b])$. We rewrite for convenience (3.5.13) as

$$(\mathbf{EVP}) \begin{cases} Lu = \lambda u & \text{in }]a,b[\\ u(a) = u(b) = 0. \end{cases} \tag{3.5.30}$$

1. The (possible) eigenvalues λ of (EVP) are all > 0

Indeed, suppose that (3.5.30) is satisfied by some $\lambda \in \mathbb{R}$ and $u \neq 0$. Then multiplying both members of the equation by $v \in E_0$ and integrating, we have

$$a(u,v) = \lambda\langle u, v\rangle \quad \forall v \in E_0, \tag{3.5.31}$$

where a is the Dirichlet form associated with L (see (3.5.17)), which is well defined also for piecewise C^1 functions. Taking in particular $v = u$ in (3.5.31) yields

$$\lambda = \frac{a(u,u)}{\int_a^b u^2 dx} = \frac{\int_a^b [pu'^2 + qu^2]\,dx}{\int_a^b u^2 dx} \equiv Q(u). \tag{3.5.32}$$

Since $p > 0$ and $q \geq 0$ in $[a,b]$, it follows that the ratio Q appearing in (3.5.32) – which is called the **Rayleigh quotient** of L – is > 0 for every $u \in E_0$, $u \neq 0$. Thus, eigenvalues of (**EVP**) are values of the Rayleigh quotient and are *a priori* > 0.

2. The key point for the existence of the eigenvalues

If the greatest lower bound of the Rayleigh quotient

$$\lambda_1 = \inf_{u\in E_1} \frac{a(u,u)}{\int_a^b u^2 dx} = \inf_{u\in E_1} Q(u), \quad E_1 = \{u \in E_0 : u \neq 0\}$$

is **attained**, then λ_1 is the smallest eigenvalue, and the minimizing functions u_1 (i. e., the functions $u_1 \in E_1$ such that $Q(u_1) = \lambda_1$) are eigenfunctions corresponding to λ_1. This goes as follows: it is clear from point **1.** that λ_1 is the smallest possible eigenvalue. Moreover, the minimum conditions defining λ_1 and u_1 can be written as

$$a(u,u) - \lambda_1\langle u,u\rangle \geq 0 \quad \forall u \in E_0 \quad \text{and} \quad a(u_1,u_1) - \lambda_1\langle u_1,u_1\rangle = 0$$

and imply that $a(u_1, v) = \lambda_1\langle u_1, v\rangle$ for all $v \in E_0$; this is a sort of Fermat theorem (Theorem 0.0.6) adapted to the present situation. Assuming in addition that $u_1 \in C^2$, this last equality is equivalent to

$$\langle Lu_1 - \lambda_1 u_1, v\rangle = 0 \quad \forall v \in E_0,$$

and from this one can conclude that $Lu_1 - \lambda_1 u_1 = 0$, which proves our statement.

Once this first step is done, repeat it but considering functions orthogonal to u_1; that is, define

$$\lambda_2 = \inf_{u\in E_2} \frac{a(u,u)}{\int_a^b u^2 dx} = \inf_{u\in E_2} Q(u), \quad E_2 = \{u \in E_0 : u \neq 0, u \perp u_1\}.$$

Then if λ_2 is attained, λ_2 is an eigenvalue of (**EVP**) and the functions $u_2 \in E_2$ such that $Q(u_2) = \lambda_2$ are eigenfunctions corresponding to λ_2. Clearly, $\lambda_2 \geq \lambda_1$, but in fact strict inequality holds, for if $\lambda_2 = \lambda_1$ we would have **two** orthogonal (and hence l. i.) eigenfunctions u_1, u_2 corresponding to the same eigenvalue, which is excluded by what we have seen in point **(a)** of the comments to Proposition 3.5.1. Therefore, $\lambda_2 > \lambda_1$, and this gives full strength to the statement that λ_2 is the *second eigenvalue* of (**EVP**).

Now iterate the procedure to find an increasing infinite sequence of eigenvalues λ_k with

$$\lambda_k = \inf_{u\in E_k} \frac{a(u,u)}{\int_a^b u^2 dx}, \quad E_k = \{u \in E_0 : u \neq 0, u \perp u_1, \dots, u_{k-1}\}.$$

The hard part in this construction consists precisely in proving that the successive infima defined above are attained; for this delicate point Weinberger refers to Courant's book [13]. We will return briefly to this in Remark 3.5.3.

3. The Green's function

Since by point **1.** the problem (**EVP**) has for $\lambda = 0$ only the solution $u = 0$, it is possible to construct (see [2]) a continuous function $G = G(s,t)$ defined in the square $[a,b] \times [a,b]$ such that the unique solution of the problem

$$\begin{cases} Lu = v & \text{in }]a,b[\\ u(a) = u(b) = 0 \end{cases} \tag{3.5.33}$$

is given by

$$u(x) = \int_a^b G(x,y)v(y)\,dy.$$

It follows that (**EVP**) is equivalent to the **integral equation**

$$u(x) = \lambda \int_a^b G(x,y)u(y)\,dy. \tag{3.5.34}$$

Using the properties of the Green's function G, Weinberger first shows that:

(a) $\lambda_n \to \infty$ as $n \to \infty$;
(b) from (a), it follows that the eigenfunctions (u_n) are complete **for functions in** E_0 (that is, the subspace spanned by them is dense in E_0), and from this, by an approximation argument similar to that used in Lemma 3.2.2, it follows that they are complete in the whole space $C([a,b])$;
(c) moreover, using again the properties of the Green's function, it also follows that for functions $u \in E_0$ – which are both more regular **and** satisfy the boundary conditions – the convergence to u of its generalized Fourier series

$$\sum_{1}^{\infty} \langle u, u_n \rangle u_n$$

is **uniform** on $[a,b]$.

4. Further properties of eigenvalues and eigenfunctions

(i) The first eigenfunction does not vanish for $a < x < b$.
(ii) **Monotonicity theorem**: The eigenvalues of (**EVP**) depend monotonically on the interval and on the coefficients; more precisely, each eigenvalue λ_k increases if the length $b - a$ of the interval $[a,b]$ decreases or if the coefficients p, q of L increase.
(iii) We have the **separation theorem** between consecutive zeros of an eigenfunction; see [2].
(iv) **Oscillation theorem**: The kth eigenfunction u_k has exactly $k - 1$ zeros in the open interval $]a,b[$.

A more systematic and comprehensive approach to the Sturm–Liouville problem and similar ones requires advanced tools from functional analysis, and in particular the framework of Sobolev spaces and operator theory. The matter is as important as technically heavy, and here again we will only indicate a few points of this approach, with the proviso that many definitions and results are given in an incomplete and possibly imprecise fashion. Our goal is mainly to invite the interested reader to further study the subject, referring for instance to the particularly clear and systematic treatment in Brezis [14].

(a) A new function space

Put

$$H^1(a,b) = \{u \in L^2(a,b) \mid u' \text{ exists and } u' \in L^2(a,b)\}. \tag{3.5.35}$$

In (3.5.35), u' denotes the **weak derivative** (or derivative *in the sense of distributions*) of u, which extends the ordinary derivative and coincides with it for differentiable functions. The inner product in $H^1(a,b)$ is defined as

$$\langle u, v\rangle = \int_a^b u'v'\,dx + \int_a^b uv\,dx. \tag{3.5.36}$$

$H^1(a,b)$ is complete with respect to the norm induced by the inner product (3.5.36) and hence is a Hilbert space. Also put

$$H_0^1(a,b) = \{u \in H^1(a,b) : u(a) = u(b) = 0\}. \tag{3.5.37}$$

$H_0^1(a,b)$ is a closed subspace of $H^1(a,b)$, and therefore is itself a Hilbert space, which will be henceforth denoted with H.

Remark: In a sense, the space $H_0^1(a,b)$ takes here the role of the space E defined in (3.5.29): it is a larger space and has the advantage of being a Hilbert space.

(b) Reduction of (EVP) to its weak form

(**EVP**) is shown to be equivalent to its *weak form*, which consists in finding a $u \in H, u \neq 0$, such that

$$a(u,v) = \lambda \int_a^b uv\,dx \quad \forall v \in H, \tag{3.5.38}$$

where a is the Dirichlet form associated with our problem, introduced in (3.5.17) and well defined also for functions in $H^1(a,b)$. Since in our assumptions a is a positive definite, symmetric, bilinear form on $H \times H$, it can be taken as a new inner product in H (inducing an equivalent norm), and we put

$$\langle u, v\rangle_1 = a(u,v) \quad (u, v \in H). \tag{3.5.39}$$

(c) Operator form of (EVP)

It is possible to introduce a linear operator $K : H \to H$ such that

$$\int_a^b uv\,dx = \langle Ku, v\rangle_1 \quad \forall u, v \in H \tag{3.5.40}$$

so that, using also (3.5.39), the problem (3.5.38) can be written

$$\langle u - \lambda Ku, v\rangle_1 = 0 \quad \forall v \in H, \tag{3.5.41}$$

that is, $u - \lambda Ku = 0$, or, putting $\mu = 1/\lambda$ (recall that $\lambda = 0$ is not an eigenvalue of (**EVP**)),

$$Ku = \mu u, \quad u \neq 0, \tag{3.5.42}$$

so that our original problem has been transformed into the **eigenvalue problem for the operator** K.

Properties of K

K is *symmetric* – that is, $\langle Ku, v\rangle_1 = \langle Kv, u\rangle_1$ for all $u, v \in H$ – as is evident from (3.5.40); moreover, K is *compact* (which is not surprising if we think of K as being essentially the integral operator induced by the Green's function [see equation (3.5.34)] and recall Example 2.7.3 in the Appendix to Chapter 2); finally, K is (strictly) *positive* in the sense that $\langle Ku, u\rangle_1 > 0 \quad \forall u \in H, u \neq 0$.

(d) The spectral theorem for compact, symmetric, positive operators in Hilbert space

Theorem 3.5.2. *Let K be a linear compact, symmetric, positive operator acting in a real infinite-dimensional Hilbert space H. Then:*

(i) *K possesses an infinite sequence (μ_k) of eigenvalues, with $\mu_k > 0$ for every k and $\mu_k \to 0$ as $k \to \infty$. Each eigenvalue μ_k has finite geometric multiplicity (i. e., the associated eigenspace is finite-dimensional) and the unit eigenvectors (u_k) corresponding to the eigenvalues form a total orthonormal system in H.*

(ii) *Moreover, the eigenvalues can be naturally arranged in a decreasing sequence,*

$$\mu_1 \geq \mu_2 \ldots\ldots \mu_k \geq \ldots,$$

where each eigenvalue is repeated as many times as its multiplicity; in addition, they enjoy (together with the corresponding eigenfunctions (u_k)) a **variational characterization** *based on the following iterative process:*

$$\mu_1 = \max_{u \in S} \langle Ku, u\rangle = \langle Ku_1, u_1\rangle, \tag{3.5.43}$$

where $S = \{u \in H : \|u\| = 1\}$ is the unit sphere in H and, for $k > 1$,

$$\mu_k = \max_{u \in S_k} \langle Ku, u\rangle = \langle Ku_k, u_k\rangle, \quad \text{where } S_k = \{u \in S : u \perp u_1, \ldots, u_{k-1}\}. \tag{3.5.44}$$

Remark 3.5.3. In the statement of Theorem 3.5.2, one relevant point comprises the fact that the successive maxima defined in (3.5.43) and (3.5.44) are achieved; this is due precisely to the compactness of the operator K and to the Hilbert structure of the underlying space.

Remark 3.5.4. More refined and complete forms of Theorem 3.5.2 – which in particular do not require the positive (or negative) definiteness of K, thus giving rise to the existence of eigenvalues of opposite sign – are to be found in Brezis [14] and also, for instance, in Dieudonné [7] and Edmunds–Evans [11]. We also note that in the literature similar results are stated and proved in the more general context of *complex* Hilbert spaces, in which case they refer to *self-adjoint* (rather than symmetric) operators.

Exercises

E1. Solutions of some of the exercises given in the text

Section 3.1

Exercise 1.1

Properties (i), (ii), and (iii) of a scalar product (Definition 3.1.1) are immediately verified using the properties of the integral, for we know that given any three functions f, g, and h in $C([a,b])$ and any $\alpha \in \mathbb{R}$ we have (omitting for simplicity the x-variable in these functions)

$$\int_a^b (f+g)h\,dx = \int_a^b (fh+gh)\,dx = \int_a^b fh\,dx + \int_a^b gh\,dx$$

$$\int_a^b \alpha fg\,dx = \alpha\int_a^b fg\,dx$$

$$\int_a^b gf\,dx = \int_a^b fg\,dx.$$

Moreover, $\int_a^b f^2\,dx \geq 0$ for every $f \in C([a,b])$, and if this integral is 0, then on the basis of Proposition 3.1.1 we have necessarily ($f^2(x) = 0$ and hence) $f(x) = 0$ for every $x \in [a,b]$. To prove Proposition 3.1.1, argue by contradiction and suppose on the contrary that $g(x_0) > 0$ for some $x_0 \in [a,b]$; then – assuming for instance that x_0 is interior to $[a,b]$ – we have $g(x) > 0$ for every x in a neighborhood $[x_0-\delta, x_0+\delta] \subset [a,b]$ of x_0. By the Weierstrass theorem (Theorem 0.0.3), we will then have $g(x) \geq K > 0$ for some K and for all $x \in [x_0-\delta, x_0+\delta]$, and therefore (by the monotonicity property of the integral of a positive function with respect to the interval of integration)

$$\int_a^b g\,dx \geq \int_{x_0-\delta}^{x_0+\delta} g\,dx \geq 2\delta K > 0,$$

contradicting the assumption that $\int_a^b g\,dx = 0$.

Exercise 1.2

For the solution of this exercise, see Proposition 2.2.1 in Chapter 2.

Exercise 1.3

The triangle property of the norm (property (iii) in Example 2.2.2) says that

$$\|x + y\| \le \|x\| + \|y\| \quad \forall x, y \in E.$$

It follows that

$$\|x\| = \|x - y + y\| \le \|x - y\| + \|y\|$$

and hence that

$$\|x\| - \|y\| \le \|x - y\| \quad \forall x, y \in E. \tag{3.5.45}$$

Interchanging the roles of x and y in (3.5.45) yields

$$\|y\| - \|x\| \le \|y - x\| = \|x - y\| \quad \forall x, y \in E. \tag{3.5.46}$$

Putting together (3.5.45) and (3.5.46) we have

$$-\|x - y\| \le \|x\| - \|y\| \le \|x - y\|,$$

which is equivalent to

$$\big|\|x\| - \|y\|\big| \le \|x - y\|.$$

This shows that the norm function is not only continuous, but *Lipschitzian* (Definition 2.2.9) of constant 1 on E.

Exercise 1.4

Recall that $c = \inf_{x \in A} f(x)$ is by definition the greatest lower bound of the set $f(A) \subset \mathbb{R}$, so that we have $f(x) \ge c$ for every $x \in A$ and

$$\forall \epsilon > 0, \quad \exists x \in A : f(x) < c + \epsilon.$$

Using these two properties with $\epsilon = \frac{1}{n}$ $(n \in \mathbb{N})$ we prove the existence of a sequence $(x_n) \subset A$ such that

$$c \le f(x_n) < c + \frac{1}{n}, \quad n \in \mathbb{N},$$

whence the conclusion follows on letting $n \to \infty$.

Section 3.2

Exercise 2.1

We have

$$\int xe^{-nx}dx = \frac{1}{n^2}\int nxe^{-nx}d(nx) = \frac{1}{n^2}\left(\int ye^{-y}dy\right)_{y=nx}.$$

However,

$$\int ye^{-y}dy = -ye^{-y} + \int e^{-y}dy = -e^{-y}(y+1) + C.$$

Therefore,

$$\int xe^{-nx}dx = -\frac{e^{-nx}}{n^2}(nx+1) + C,$$

so that finally

$$\int_0^1 n^2xe^{-nx}dx = -\left[e^{-nx}(nx+1)\right]_{x=0}^{x=1} = -\left[e^{-n}(n+1) - 1\right] \to 1 \quad (n \to \infty).$$

Exercise 2.2

Example 3.2.4: $f(x) = x$ ($|x| \le \pi$)

As f is an odd function, the coefficients $a_0 = a_k = 0 \quad \forall k \in \mathbb{N}$, while

$$b_k = \frac{1}{\pi}\int_{-\pi}^{\pi} f(x)\sin kx\,dx = \frac{2}{\pi}\int_0^{\pi} x\sin kx\,dx.$$

However, for $k \in \mathbb{N}$,

$$\int_0^{\pi} x\sin kx\,dx = \left[-\frac{x\cos kx}{k}\right]_{x=0}^{x=\pi} + \int_0^{\pi}\frac{\cos kx}{k}dx$$
$$= -\frac{\pi\cos k\pi}{k} + \frac{1}{k}\left[\frac{\sin kx}{k}\right]_{x=0}^{x=\pi} = -\frac{\pi}{k}(-1)^k = (-1)^{k+1}\frac{\pi}{k}.$$

Hence,

$$b_k = \frac{2}{\pi}(-1)^{k+1}\frac{\pi}{k} = (-1)^{k+1}\frac{2}{k}.$$

Example 3.2.6: $f(x) = |x|$ $(|x| \leq \pi)$
As f is an even function, the coefficients $b_k = 0 \quad \forall k \in \mathbb{N}$, while

$$a_0 = \frac{1}{2\pi} \int_{-\pi}^{\pi} f(x)dx = \frac{1}{\pi} \int_{0}^{\pi} x\,dx = \frac{\pi}{2}$$

and, for $k \in \mathbb{N}$,

$$a_k = \frac{1}{\pi} \int_{-\pi}^{\pi} f(x) \cos kx\,dx = \frac{2}{\pi} \int_{0}^{\pi} x \cos kx\,dx.$$

However, for $k \in \mathbb{N}$,

$$\int_{0}^{\pi} x \cos kx\,dx = \left[\frac{x \sin kx}{k} \right]_{x=0}^{x=\pi} - \int_{0}^{\pi} \frac{\sin kx}{k}\,dx$$
$$= \frac{1}{k} \left[\frac{\cos kx}{k} \right]_{x=0}^{x=\pi} = \frac{1}{k^2}[(-1)^k - 1].$$

Hence,

$$a_k = \frac{2}{\pi} \frac{1}{k^2}[(-1)^k - 1] = -\frac{2}{k^2\pi}[1 - (-1)^k].$$

Section 3.3
Exercise 3.1
The key point for this exercise consists simply in recalling the basic fact that if a series $\sum_{n=1}^{\infty} a_n$ converges, then $a_n \to 0$ as $n \to \infty$, so that in particular the sequence (a_n) is **bounded**. Note that these assertions hold for series of vectors in any normed space (Definition 2.4.2), and not only for series with real entries. This property of convergent series immediately proves the inclusion $l^p \subset l^\infty$. Also to prove that $l^1 \subset l^p$, take $x = (x_n) \in l^1$; write

$$|x_n|^p = |x_n|^{p-1}|x_n| \leq C|x_n|$$

for some positive constant C and use the comparison criterion for series with non-negative terms to reach the conclusion that $x \in l^p$.

The same argument proves that more generally, if $1 \leq p, q \leq \infty$ and $p < q$, then $l^p \subset l^q$: just write (for $x = (x_n)$)

$$|x_n|^q = |x_n|^{q-p}|x_n|^p \leq C|x_n|^p$$

to conclude by comparison that the convergence of the series $\sum_{n=1}^{\infty} |x_n|^q$ is implied by that of the series $\sum_{n=1}^{\infty} |x_n|^p$.

Of course, the same conclusions can be reached observing that, if $p < q$, then

$$|x_n|^q \leq |x_n|^p$$

for a given $x = (x_n)$ and sufficiently large n, because $x_n \to 0$ as $n \to \infty$, implying that $|x_n| < 1$ for large n.

The interest of this exercise lies also in the fact that the situation can be quite different for the general spaces L^p, as far as inclusions between them are concerned; see Remark 3.4.2 (and consider that in the present situation the measure is *not* finite).

Exercise 3.2

This is a direct application of formula (3.1.12),

$$y = \sum_{i=1}^{n} \langle x, e_i \rangle e_i,$$

giving (in any inner product space E) the orthogonal projection y of a vector $x \in E$ onto the subspace M spanned by n orthonormal vectors $e_1, \dots, e_n$ of E. Indeed, if $a = (a_n)_{n\in\mathbb{N}} \in E = l^2$ – equipped with the scalar product (3.3.13) – and for each $n \in \mathbb{N}$, e_n is defined via

$$e_n = (0, 0, \dots, 0, 1, 0, \dots),$$

where the number 1 appears only at the nth place of the sequence, then

$$\langle a, e_i \rangle = a_i \quad \forall i \in \mathbb{N}$$

so that

$$\begin{aligned}
\sum_{i=1}^{k} \langle a, e_i \rangle e_i &= a_1 e_1 + \cdots a_k e_k \\
&= a_1(1, 0, 0, \dots) + a_2(0, 1, 0, \dots) + \cdots + a_k(0, 0, \dots, 1, 0 \dots) \\
&= (a_1, a_2, \dots, a_k, 0, 0 \dots).
\end{aligned}$$

Section 3.4

Exercise 4.1

The relations

$$\int_{-\infty}^{+\infty} \frac{1}{\sqrt{1 + x^2}}\, dx = +\infty, \quad \int_{-\infty}^{+\infty} \frac{1}{1 + x^2}\, dx = \pi$$

show an $f \in L^2(\mathbb{R})$ which does not belong to $L^1(\mathbb{R})$.

Exercise 4.2

Put $f(x) = \langle x, a\rangle$ for $x \in E$. Given a convergent series $\sum_{n=1}^{\infty} x_n$ of vectors of E, put

$$s_k = \sum_{n=1}^{k} x_n \quad (k \in \mathbb{N}), \quad s = \sum_{n=1}^{\infty} x_n \tag{3.5.47}$$

so that $s_k \to s$ in E as $k \to \infty$ by definition. Since

$$f(s_k) = \langle s_k, a\rangle = \sum_{n=1}^{k} \langle x_n, a\rangle \quad \forall k \in \mathbb{N}, \tag{3.5.48}$$

letting $k \to \infty$ and using the continuity of f on E we see that the numerical series $\sum_{n=1}^{\infty} \langle x_n, a\rangle$ converges and its sum is $f(s)$, as desired.

Exercise 4.3

Let x, y belong to the closure $\overline{M}$ of the vector subspace M of a normed space E. To show that $x + y \in \overline{M}$, let $(x_n), (y_n)$ be sequences in M such that $x_n \to x, y_n \to y$ as $n \to \infty$. Then $x_n + y_n \to x + y$, as follows from the inequality

$$\|(x_n + y_n) - (x + y)\| \le \|x_n - x\| + \|y_n - y\|.$$

Since $x_n + y_n \in M$ for every n, it follows that $x + y \in \overline{M}$. Similarly, one checks that $ax \in \overline{M}$ if $x \in \overline{M}$ and $a \in \mathbb{R}$.

Exercise 4.4

Let $F(x) = (\langle x, e_n\rangle)_{n\in\mathbb{N}}$, where $x \in H$ and (e_n) is orthonormal. Then:

- $F(x) \in l^2$ by virtue of Bessel's inequality:

$$\|F(x)\|_{l^2} = \sum_{n=1}^{\infty} |\langle x, e_n\rangle|^2 \le \|x\|^2. \tag{3.5.49}$$

- For $x, y \in H$ we have, by the properties of the inner product and the definition of sum of two sequences,

$$F(x + y) = (\langle x + y, e_n\rangle)_{n\in\mathbb{N}} = (\langle x, e_n\rangle)_{n\in\mathbb{N}} + (\langle y, e_n\rangle)_{n\in\mathbb{N}} = F(x) + F(y),$$

  so that F is continuous (F is a bounded linear operator of H into l^2 in the sense of Example 2.2.7).
- Saying that F is surjective means that for every $a = (a_n)_{n\in\mathbb{N}} \in l^2$, there is an $x \in H$ such that $F(x) = (\langle x, e_n\rangle)_{n\in\mathbb{N}} = a$, i. e., $\langle x, e_n\rangle = a_n \quad \forall n \in \mathbb{N}$. As shown in the proof of the Fischer–Riesz theorem, this holds taking

$$x = \sum_{n=1}^{\infty} a_n e_n.$$

- Suppose in addition that the orthonormal system is **total**. Then if $F(x) = (\langle x, e_n \rangle)_{n\in\mathbb{N}} = 0$ (that is, if $\langle x, e_n \rangle = 0\ \forall n \in \mathbb{N}$), it follows by the expansion $x = \sum_{n=1}^{\infty} \langle x, e_n \rangle e_n$ (Corollary 3.4.4) that $x = 0$. Thus, in this case F is also injective.

4 Partial differential equations

Introduction

The purpose of this chapter is that of accompanying the student in his/her very first steps into the extremely vast field of PDEs. While Chapter 1 of this book could be taken as a brief *résumé* of some relevant facts concerning ODEs (at least for the practice if not for the theory), this cannot be the case for the present chapter about PDEs, the reason being the elementary nature of the present book on one side and the intrinsic complexity and variety of PDEs themselves on the other. To have an idea of this complexity, just consider that the relatively narrow field of linear equations of second order with constant coefficients (which for ODEs can be treated on one page) needs the fundamental and classical separation into elliptic, parabolic, and hyperbolic equations even to formulate local existence results for the Cauchy problem for them.

Thus, we shall here restrict ourselves to presenting some classical examples of PDEs, which indeed many science students meet in one form or another mainly from physics. For instance, we learn from the study of electromagnetism that the electric field $\boldsymbol{E} = \boldsymbol{E}(x, y, z)$ satisfies the law

$$\operatorname{div} \boldsymbol{E} = \rho, \tag{4.0.1}$$

where $\rho = \rho(x, y, z)$ is the **electric charge density** which is distributed in a region Ω of the physical space $\mathbb{R}^3$. On the other hand, $\boldsymbol{E}$ is a **conservative** field, which means that it comes from an electric **potential** ϕ:

$$\boldsymbol{E} = \operatorname{grad} \phi = \nabla \phi. \tag{4.0.2}$$

Since $\operatorname{div} \boldsymbol{E} = \frac{\partial E_1}{\partial x} + \frac{\partial E_2}{\partial y} + \frac{\partial E_3}{\partial z}$ (where E_1, E_2, E_3 are the components of $\boldsymbol{E}$), (4.0.1) and (4.0.2) give the equation for the electric potential:

$$\frac{\partial^2 \phi}{\partial x^2} + \frac{\partial^2 \phi}{\partial y^2} + \frac{\partial^2 \phi}{\partial z^2} = \rho. \tag{4.0.3}$$

This is the famous **Poisson equation**, which must be solved if we want to know the spatial distribution in Ω of the electric potential ϕ. If our electrically conducting body $\Omega \subset \mathbb{R}^3$ is bounded by a surface $\partial\Omega$, it is usually required that ϕ be held **fixed** at a certain level on $\partial\Omega$:

$$\phi_{|\partial\Omega} = \text{const} = K, \tag{4.0.4}$$

or more generally, that it equals an assigned function ϕ_0 on $\partial\Omega$. Equations (4.0.3) and (4.0.4) constitute one of the classical BVPs of mathematical physics, and solving it requires finding a function that is regular enough (C^2) in Ω to satisfy the differential equation (4.0.3) and takes on **with continuity** its value K on the limiting surface $\partial\Omega$.

https://doi.org/10.1515/9783111302522-004

The difficulty in solving this problem depends very much also on the geometry of Ω, and a solution can be constructed by the elementary methods of **separation of variables** in case Ω is a **ball**, so that $\partial\Omega$ is a spherical surface; this leads in particular to the study of the spherical harmonics, as is well explained in Weinberger's classic book [2], which contains a rich amount of information on the entire field of linear second-order PDEs.

A good part (Sections 4.2 and 4.4) of this chapter, whose level is definitely elementary, is devoted to discussing the simplest case of this problem, in which we have only two variables and Ω is a rectangle. The deliberate reduction of the technical and computational complexity should hopefully permit a better understanding of the method of separation of variables and eigenfunction expansion (which includes an additional study of Fourier series, see Section 4.3) and thus invite to a further and technically more adequate study of this and companion problems.

Section 4.1 of this chapter presents a few generalities and some simple examples of PDEs, paving the way to introduce the BVPs and IVPs of mathematical physics involving the Poisson, heat, and wave equations. Mention is made of Gauss' theorem and an example of its use to prove uniqueness in some of the cited problems is given.

The final Section 4.5 aims at testing the utility and flexibility of the method seen in Sections 4.2 and 4.4 in order to deal with the heat equation. We try to make it clear that a key point for this approach is the spectral theorem for the Laplacian, giving rise to a complete orthonormal system of eigenfunctions along with Courant–Hilbert's famous result [13].

Classical references for the study of PDEs are, among others, the books of Lions [15], Trèves [16], and Gilbarg and Trudinger [17]. To these, we add the more recent one by Brezis [14] recommending it for his high clarity and organization in this complex, vast, and delicate matter.

4.1 Partial differential equations (PDEs). The PDEs of mathematical physics

PDEs are differential equations in which the unknown function depends upon **several** variables (rather than just one, as in the case of ODEs); hence, they will contain – besides the independent variables – the unknown function and its **partial derivatives** up to some order, which will be the **order of the PDE**.

Recall that the general form of a first-order ODE is

$$F(x, u, u') = 0, \tag{4.1.1}$$

though usually one considers the special "normal form" of (4.1.1),

$$u' = f(x, u).$$

The general form of a first-order PDE **in two variables** is

$$F\left(x, y, u, \frac{\partial u}{\partial x}, \frac{\partial u}{\partial y}\right) = 0, \tag{4.1.2}$$

where F is a given real-valued function defined in a set $D \subset \mathbb{R}^5$. Note that there is no “normal form” for PDEs. A **solution** of (4.1.2) is a real-valued function u defined in a (usually open and connected) subset $A \subset \mathbb{R}^2$ that satisfies **pointwisely** (4.1.2) in A, that is, it is such that

$$F\left(x, y, u(x,y), \frac{\partial u}{\partial x}(x,y), \frac{\partial u}{\partial y}(x,y)\right) = 0 \quad \forall (x,y) \in A; \tag{4.1.3}$$

of course, one must pre-require that u has partial derivatives u_x, u_y at every point of A and that the argument $(x, y, u(x,y), u_x(x,y), u_y(x,y))$ of F appearing in (4.1.3) belongs to its domain of definition D for every $(x,y) \in A$. As a matter of fact, it is usually required that $u \in C^1(A)$.

Example 4.1.1. Consider the equation

$$y\frac{\partial u}{\partial x} - x\frac{\partial u}{\partial y} = 0. \tag{4.1.4}$$

Here we have $F(x, y, u, p, q) = yp - xq$, and the function $u(x,y) = x^2 + y^2$ is a solution of (4.1.4). This is an example of a **linear** first-order equation, the general form of these being as follows:

$$a(x,y)\frac{\partial u}{\partial x} + b(x,y)\frac{\partial u}{\partial y} + c(x,y)u = d(x,y), \tag{4.1.5}$$

where the **coefficient functions** a, b, c, d are defined in some $A_0 \subset \mathbb{R}^2$, so that – with reference to (4.1.2) – here the function F has the form

$$F(x, y, u, p, q) = a(x,y)p + b(x,y)q + c(x,y)u - d(x,y)$$

and is defined on $D = A_0 \times \mathbb{R}^3$.

On the other hand, the equation

$$\frac{\partial u}{\partial x} + u\frac{\partial u}{\partial y} = 0$$

is a simple example of a **nonlinear** first-order PDE.

PDEs of the second order

A second-order PDE in two variables will have the form

$$F\left(x, y, u, \frac{\partial u}{\partial x}, \frac{\partial u}{\partial y}, \frac{\partial^2 u}{\partial x^2}, \frac{\partial^2 u}{\partial x \partial y}, \frac{\partial^2 u}{\partial y^2}\right) = 0. \tag{4.1.6}$$

Here $F = F(x, y, u, p, q, r, s, t) : D \subset \mathbb{R}^8 \to \mathbb{R}$. The concept of solution will be given accordingly as a function $u \in C^2(A)$, with $A \subset \mathbb{R}^2$ open and connected, satisfying pointwisely (4.1.6).

Example 4.1.2. Consider the following equations:

$$\frac{\partial^2 u}{\partial x^2} + \frac{\partial^2 u}{\partial y^2} = f(x, y) \quad \text{(Poisson equation)}, \tag{4.1.7}$$

$$\frac{\partial^2 u}{\partial x^2}\frac{\partial^2 u}{\partial y^2} - \left(\frac{\partial^2 u}{\partial x \partial y}\right)^2 = f(x, y) \quad \text{(Monge–Ampère equation)}, \tag{4.1.8}$$

or

$$\det H(u) = u_{xx}u_{yy} - (u_{xy})^2 = f(x, y).$$

Formula (4.1.7) is a **linear** equation, while (4.1.8) is **nonlinear**.

If $f \equiv 0$, then

$$u(x, y) = \ln\sqrt{x^2 + y^2}, \quad (x, y) \neq (0, 0), \tag{4.1.9}$$

is a solution of the first. This example shows that, unlike the case $n = 1$, the solutions of linear equations with constant coefficients are not defined everywhere.

Of course one can write these and similar equations in more than two variables. For instance, the three-dimensional version of (4.1.7) is

$$\frac{\partial^2 u}{\partial x^2} + \frac{\partial^2 u}{\partial y^2} + \frac{\partial^2 u}{\partial z^2} = f(x, y, z) \tag{4.1.10}$$

and for $f = 0$ has the elementary solution

$$u(x, y, z) = \frac{1}{\sqrt{x^2 + y^2 + z^2}} = \frac{1}{\|\mathbf{x}\|}, \quad \mathbf{x} \neq \mathbf{0}. \tag{4.1.11}$$

Exercise 4.1.1. Check that the functions u given in (4.1.9) and (4.1.11) are solutions of the homogeneous Poisson equation (which is named the **Laplace equation**) in two and three variables, respectively. Apart from a constant factor, they are the so-called **fundamental solutions** of the Laplace equation.

Explicit resolution of PDEs

Finding the "general solution" of a PDE, that is, finding all of its solutions, can be done only in a few very simple cases. For instance, the first-order PDEs

$$\frac{\partial u}{\partial x}=0, \quad \frac{\partial u}{\partial y}=0 \tag{4.1.12}$$

have respectively as solutions

$$u(x,y)=c(y), \quad u(x,y)=d(x), \tag{4.1.13}$$

where c and d are arbitrary functions.

Examples. Consider the following equations and their solutions:

$$\frac{\partial u}{\partial x}-\frac{\partial u}{\partial t}=0, \quad u(x,t)=f(x+t) \quad (f \text{ arbitrary}), \tag{4.1.14}$$

$$\frac{\partial^2 u}{\partial x^2}=0, \quad u(x,y)=c(y)x+d(y) \quad (c,d \text{ arbitrary}), \tag{4.1.15}$$

$$\frac{\partial^2 u}{\partial x \partial y}=0, \quad u(x,y)=f(x)+g(y) \quad (f,g \text{ arbitrary}), \tag{4.1.16}$$

$$\frac{\partial^2 u}{\partial t^2}-v^2\frac{\partial^2 u}{\partial x^2}=0, \quad u(x,t)=f(x+tv)+g(x-tv) \quad (f,g \text{ arbitrary}). \tag{4.1.17}$$

To solve (4.1.14), make the change of variables

$$\begin{cases} x+t=p \\ x-t=q. \end{cases}$$

Let $\omega:\mathbb{R}^2\to\mathbb{R}$ be such that

$$u(x,t)=\omega(p,q)=\omega(x+t,x-t). \tag{4.1.18}$$

Then by the chain rule we have

$$\begin{cases} \frac{\partial u}{\partial x} = \frac{\partial \omega}{\partial p}\frac{\partial p}{\partial x}+\frac{\partial \omega}{\partial q}\frac{\partial q}{\partial x} = \frac{\partial \omega}{\partial p}+\frac{\partial \omega}{\partial q} \\ \frac{\partial u}{\partial t} = \frac{\partial \omega}{\partial p}\frac{\partial p}{\partial t}+\frac{\partial \omega}{\partial q}\frac{\partial q}{\partial t} = \frac{\partial \omega}{\partial p}-\frac{\partial \omega}{\partial q} \end{cases}$$

so that

$$\frac{\partial u}{\partial x}-\frac{\partial u}{\partial t}=2\frac{\partial \omega}{\partial q},$$

showing that in the new variables p,q equation (4.1.14) is transformed to

$$\frac{\partial \omega}{\partial q}=0,$$

whence we get $\omega(p,q) = f(p)$ with arbitrary f, and finally returning to the old variables – that is, using (4.1.18) – we conclude that

$$u(x,t) = f(x+t).$$

As to (4.1.15), from $\frac{\partial^2 u}{\partial x^2} = \frac{\partial}{\partial x}(\frac{\partial u}{\partial x}) = 0$ we first obtain

$$\frac{\partial u}{\partial x} = c(y),$$

and integrating again with respect to x, it follows that $u(x,y) = c(y)x + d(y)$.

Likewise for (4.1.16), from $\frac{\partial^2 u}{\partial x \partial y} = \frac{\partial}{\partial x}(\frac{\partial u}{\partial y}) = 0$ we first obtain

$$\frac{\partial u}{\partial y} = c(y),$$

whence integrating with respect to y, we have $u(x,y) = C(y) + D(x)$ with C a primitive of c.

Exercise 4.1.2. The solution of (4.1.17) is obtained making essentially the same change of variables used for (4.1.14), or more precisely, putting $x + tv = p, x - tv = q$. Also check that any u of the form written in (4.1.17) – with $f, g : \mathbb{R} \to \mathbb{R}$ twice differentiable, of course – satisfies the equation.

Linear PDEs of the second order

A linear PDE of the second order (in two variables) has the form

$$\begin{cases} a(x,y)\frac{\partial^2 u}{\partial x^2} + 2b(x,y)\frac{\partial^2 u}{\partial x \partial y} + c(x,y)\frac{\partial^2 u}{\partial y^2} + \\ \quad + d(x,y)\frac{\partial u}{\partial x} + e(x,y)\frac{\partial u}{\partial y} + f(x,y)u = g(x,y), \end{cases} \tag{4.1.19}$$

where the **coefficient functions** $a, b, c \dots, g$ are defined and continuous in an open subset A_0 of $\mathbb{R}^2$. A special case of (4.1.19) are the linear PDEs with **constant coefficients**:

$$a\frac{\partial^2 u}{\partial x^2} + 2b\frac{\partial^2 u}{\partial x \partial y} + c\frac{\partial^2 u}{\partial y^2} + +d\frac{\partial u}{\partial x} + e\frac{\partial u}{\partial y} + fu = g. \tag{4.1.20}$$

The **classification** of linear second-order PDEs with constant coefficients is made on the basis of the top order coefficients a, b, c appearing in (4.1.20) as follows:

$$\begin{cases} ac - b^2 > 0 & \textbf{elliptic} \\ ac - b^2 = 0 & \textbf{parabolic} \\ ac - b^2 < 0 & \textbf{hyperbolic}. \end{cases} \tag{4.1.21}$$

Examples of equations falling in each of the three classes are:

- the **Poisson** equation,

$$\frac{\partial^2 u}{\partial x^2} + \frac{\partial^2 u}{\partial y^2} = g, \tag{4.1.22}$$

- the **heat** equation,

$$\frac{\partial u}{\partial t} - \frac{\partial^2 u}{\partial x^2} = g, \tag{4.1.23}$$

- the **wave** equation,

$$\frac{\partial^2 u}{\partial t^2} - \frac{\partial^2 u}{\partial x^2} = g. \tag{4.1.24}$$

These three equations are usually called **the PDEs of mathematical physics.**

We shall study the equations of mathematical physics accompanied by **supplementary conditions** upon the unknown function u and/or its first-order partial derivatives.

These problems will have the following common structure:

$$\begin{cases} L[u] = f \\ \text{BC/IC}, \end{cases} \tag{4.1.25}$$

where in the first row we have the PDE itself, represented by a **linear differential operator** L acting on the function u (f is given), and in the second row the term **BC/IC** stands for **boundary conditions** and/or **initial conditions**, the last appearing if the **time variable** appears in the equation.

Specific examples now follow.

A. The Poisson equation

Consider the BVP

$$\begin{cases} \Delta u \equiv \frac{\partial^2 u}{\partial x^2} + \frac{\partial^2 u}{\partial y^2} = f(x,y) & (x,y) \in \Omega \\ u = g(x,y) & (x,y) \in \partial\Omega, \end{cases} \tag{4.1.26}$$

where Ω is a bounded open set in $\mathbb{R}^2$ with boundary $\partial\Omega$, while $f : \Omega \to \mathbb{R}$ and $g : \partial\Omega \to \mathbb{R}$ are given continuous functions. The set Ω and the functions f, g are the **data**, while u is the **unknown** of the problem (4.1.26). This is called the **Dirichlet problem** for the Poisson equation and consists in finding a function $u : \overline{\Omega} = \Omega \cup \partial\Omega \to \mathbb{R}$ such that

$$u \in C^2(\Omega) \cap C(\overline{\Omega})$$

and u satisfies pointwisely both the equation and the boundary condition in (4.1.26). It will be usually written more briefly as

$$\begin{cases} \frac{\partial^2 u}{\partial x^2} + \frac{\partial^2 u}{\partial y^2} = f & \text{in } \Omega, \\ u = g & \text{on } \partial\Omega. \end{cases} \tag{4.1.27}$$

Another kind of BVP for the Poisson equation is the **Neumann problem**, which is written

$$\begin{cases} \frac{\partial^2 u}{\partial x^2} + \frac{\partial^2 u}{\partial y^2} = f & \text{in } \Omega \\ \frac{\partial u}{\partial \nu} = g & \text{on } \partial\Omega \end{cases} \tag{4.1.28}$$

and consists in finding a function $u : \overline{\Omega} = \Omega \cup \partial\Omega \to \mathbb{R}$ such that

$$u \in C^2(\Omega) \cap C^1(\overline{\Omega})$$

and u satisfies pointwisely both the equation and the boundary condition in (4.1.28). The latter is given on the **outer normal derivative** $\frac{\partial u}{\partial \nu}$ of u on $\partial\Omega$, that is, the directional derivative of u in the direction of the normal unit vector $\boldsymbol{\nu}$ on $\partial\Omega$ pointing to the exterior of Ω. Some remarks follow.

- The same problems can be considered in three variables ($\Omega \subset \mathbb{R}^3$) and in fact in any number n of variables ($\Omega \subset \mathbb{R}^n$).
- For the Neumann problem (4.1.28), it is necessary to assume that $\partial\Omega$ is a **regular curve** (if $n = 2$) or a **regular surface** (if $n = 3$); this ensures that for any $\mathbf{x} \in \partial\Omega$, there is the **tangent line** (resp. the **tangent plane**) to $\partial\Omega$ at the point $\mathbf{x}$, and therefore also the **normal unit vector** $\nu(\mathbf{x})$ to $\partial\Omega$ at the point $\mathbf{x}$. Moreover, the solution u is required to belong to $C^1(\overline{\Omega})$, in such a way that the normal derivative of u,

  $$\frac{\partial u}{\partial \boldsymbol{\nu}}(\mathbf{x}) = \nabla u(\mathbf{x}) \cdot \boldsymbol{\nu}(\mathbf{x}),$$

  exists at every point $\mathbf{x} \in \partial\Omega$.
- Given any open $\Omega \subset \mathbb{R}^n$, while the definition of $C^1(\Omega)$ (and more generally of $C^k(\Omega)$) is clear, e. g.,

  $$C^1(\Omega) = \left\{ u \in C(\Omega) : \forall i : 1, \dots, n \ \frac{\partial u}{\partial x_i} \text{ (exists and) belongs to } C(\Omega) \right\},$$

  the definition of $C^1(\overline{\Omega})$ requires some explanation. We put

  $$C^1(\overline{\Omega}) = \left\{ u \in C^1(\Omega) : u, \frac{\partial u}{\partial x_i} (i = 1, \dots, n) \text{ can be extended by continuity to } \overline{\Omega} \right\}.$$

Thus, we require that for every $\mathbf{x} \in \partial\Omega$, the limits

$$\lim_{y \to x, y \in \Omega} \frac{\partial u}{\partial x_i}(\mathbf{y}) \quad (i = 1, \dots, n) \tag{4.1.29}$$

exist and are finite, and we **define** $\frac{\partial u}{\partial x_i}(\mathbf{x})$ via the limit in (4.1.29). Note that if $n = 1$ and $\Omega =]a, b[$, we have two possible definitions of derivative at the boundary points:

$$\text{(i) } u'(a) = \lim_{t \to a^+} \frac{u(t) - u(a)}{t - a} \quad \text{or} \quad \text{(ii) } u'(a) = \lim_{t \to a^+} u'(t)$$

(provided of course that the above limits are finite). However, using the L'Hopital rule, we see that if the limit in (ii) exists, then also the limit in (i) exists and they are equal.

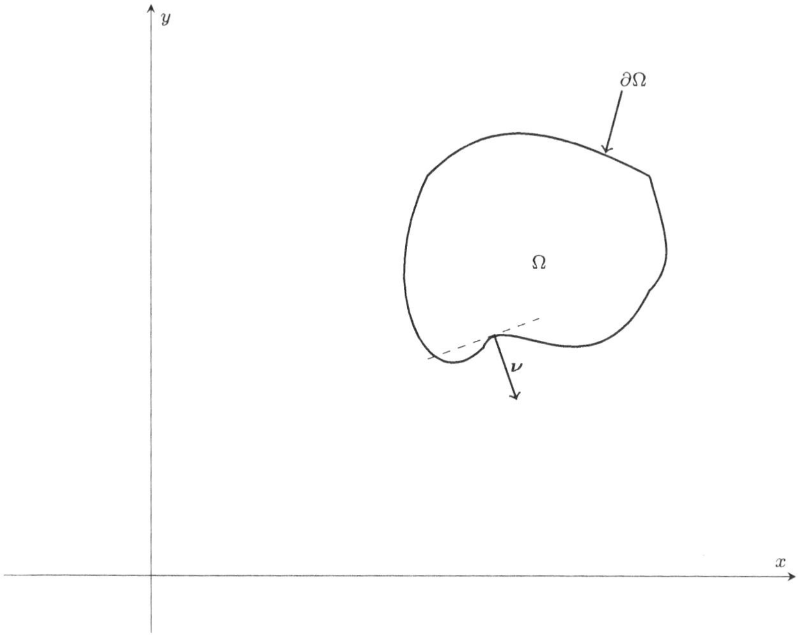

B. The heat equation

This is the equation

$$\frac{\partial u}{\partial t} - \Delta_x u = f(\mathbf{x}, t) \quad (\mathbf{x}, t) \in \Omega \times]0, +\infty[, \tag{4.1.30}$$

where t is the time variable, while Δ_x stands for the Laplacian in the space variable $\mathbf{x} = (x_1, \dots, x_n)$:

$$\Delta_x = \sum_{i=1}^{n} \frac{\partial^2 u}{\partial x_i^2}.$$

If $n = 1, 2, 3$, $u(\mathbf{x}, t)$ represents the **temperature** at the point $\mathbf{x}$ and time t in a (one-, two-, or three-dimensional) body Ω.

Problem. Determine the temperature in the body Ω knowing:
(a) the **initial temperature** ($t = 0$) of Ω and
(b) the **temperature on the boundary** $\partial\Omega$ for all $t > 0$:

$$\begin{cases} u(\mathbf{x}, 0) = u_0(\mathbf{x}) & \mathbf{x} \in \Omega \quad \textbf{(initial condition)} \\ u(\mathbf{x}, t) = g(\mathbf{x}, t) & \mathbf{x} \in \partial\Omega, t > 0 \quad \textbf{(boundary condition)}. \end{cases} \tag{4.1.31}$$

Equation (4.1.30), accompanied by the conditions in (4.1.31), is called the **Cauchy–Dirichlet problem** for the heat equation.
We can change condition (b) above to the following: (b')

$$\frac{\partial u}{\partial \nu}(\mathbf{x}, t) = h(\mathbf{x}, t) \quad \mathbf{x} \in \partial\Omega, \ t > 0,$$

thus obtaining the **Cauchy–Neumann problem** for the heat equation.

C. The wave equation

This is the equation

$$\frac{\partial^2 u}{\partial t^2} - \Delta_x u = f \quad \text{in } \Omega \times]0, +\infty[. \tag{4.1.32}$$

For $n = 2$ we have the equation of the **vibrating membrane**, in which $u(\mathbf{x}, t) = u(x, y, t)$ represents the vertical displacement ("displacement along the z-axis") at time t of that point of the membrane Ω in the plane that at rest has coordinates (x, y):

$$(x, y, 0) \to (x, y, u(x, y, t)).$$

We can in particular consider the following IVP/BVP:

$$\begin{cases} \frac{\partial^2 u}{\partial t^2} - (\frac{\partial^2 u}{\partial x^2} + \frac{\partial^2 u}{\partial y^2}) = f(x, y, t) & (x, y) \in \Omega, \ t > 0 \\ u(x, y, 0) = u_0(x, y) & (x, y) \in \Omega \\ u_t(x, y, 0) = v_0(x, y) & (x, y) \in \Omega \\ u(x, y, t) = 0 & (x, y) \in \partial\Omega, \ t > 0, \end{cases} \tag{4.1.33}$$

in which u_0 and v_0 are the initial position and velocity of the membrane (**initial conditions**), while the last condition means that the boundary of the membrane is being held fixed in the plane $z = 0$ for all time $t > 0$ (**boundary condition**).

General features of the BVP/IVP under study

Before studying the various problems presented above, it is convenient to make some general remarks valid for any of these problems; these remarks are all based on the circumstance that they are all **linear** problems. To see this in some detail, consider again (4.1.25), which we rewrite in the form

$$\begin{cases} L[u] = f & \textbf{(equation)} \\ B[u] = g & \textbf{(boundary and/or initial conditions)}. \end{cases} \tag{4.1.34}$$

Specifically, both the differential operator L and the boundary value/initial value operator B appearing in (4.1.34) are linear operators when considered among the appropriate function spaces acting as domain and codomain.

Common problems to be considered about (4.1.34) include:

- **existence** of a solution;
- **uniqueness** of the solution;
- **continuous dependence** of the solution u upon the data f and g.

Three general remarks

1. (Uniqueness)

If the homogeneous problem

$$\begin{cases} L[u] = 0 & \text{(homogeneous equation)} \\ B[u] = 0 & \text{(homogeneous BC and/or IC)} \end{cases} \tag{4.1.35}$$

has **only** the solution $u = 0$, then the inhomogeneous problem (4.1.34) has **at most one** solution. More generally, if

$$S \equiv \{u : u \text{ is a solution of (4.1.34)}\},$$
$$S_0 \equiv \{u : u \text{ is a solution of (4.1.35)}\},$$

and z is a (possible) solution of (4.1.34) – that is, $z \in S$ – then

$$S = S_0 + z.$$

The statements above are proved in exactly the same way as we did when discussing linear ODEs and systems, on relating the solution set of the “complete” (i. e., inhomogeneous) equation with that of the homogeneous one; see Lemma 1.3.2.

2. (“Superposition principle”)

To illustrate this, suppose for instance that u_1 is a solution of the problem

$$\begin{cases} L[u] = f \\ B[u] = 0 \end{cases} \tag{4.1.36}$$

and that u_2 is a solution of the problem

$$\begin{cases} L[u] = 0 \\ B[u] = g. \end{cases} \tag{4.1.37}$$

Then $u = u_1 + u_2$ is a solution of (4.1.34) (the solution of (4.1.34) if **1.** holds).

3. (Series solution of (4.1.34)**)**
Consider again the partially homogeneous problem (4.1.37) and suppose that we have a sequence (u_n) of "elementary" solutions of $L[u] = 0$. Then for any choice of the coefficients c_n, the superposition

$$u = \sum_{n=1}^{\infty} c_n u_n \tag{4.1.38}$$

will be itself a solution of (4.1.37), provided that the convergence problems involved in the series are overcome, which means checking not only that u is well defined via (4.1.38), but also the validity of the desired equality

$$L[u] = \sum_{n=1}^{\infty} c_n L[u_n] = 0.$$

Likewise, if the equality

$$B[u] = \sum_{n=1}^{\infty} c_n B[u_n]$$

is meaningful and permitted, then we see that the coefficients (c_n) must be chosen so that the sum of the series above is equal to g, in order to satisfy the side condition in (4.1.37) and hence to have a solution of that problem (the solution if **1.** holds).

The Dirichlet and Neumann problems for Poisson's equation
Consider the problems

$$(\mathbf{D}) \begin{cases} \Delta u = f & \text{in } \Omega \\ u = g & \text{on } \partial\Omega, \end{cases} \qquad (\mathbf{N}) \begin{cases} \Delta u = f & \text{in } \Omega \\ \frac{\partial u}{\partial \nu} = g & \text{on } \partial\Omega. \end{cases} \tag{4.1.39}$$

We assume that:
- Ω is an **open, bounded, connected** subset of $\mathbb{R}^n (n = 2, 3)$.
- $f : \Omega \to \mathbb{R}$ and $g : \partial\Omega \to \mathbb{R}$ are given **continuous** functions.
- Moreover, for problem (**N**), we assume that the boundary $\partial\Omega$ is a **regular curve or surface** (depending on whether $n = 2$ or $n = 3$, resp.).

A **solution** of (**D**) is a function

$$u \in C^2(\Omega) \cap C(\overline{\Omega})$$

satisfying pointwisely both the equation in Ω and the boundary condition on $\partial\Omega$. Similarly, a solution of (**N**) is a function

$$u \in C^2(\Omega) \cap C^1(\overline{\Omega})$$

satisfying pointwisely both the equation in Ω and the boundary condition on $\partial\Omega$.

On the basis of "General Remark" **1.**, we study the homogeneous problems associated with (**D**) and (**N**), namely,

$$(\mathbf{D0}) \begin{cases} \Delta u = 0 & \text{in } \Omega \\ u = 0 & \text{on } \partial\Omega, \end{cases} \qquad (\mathbf{N0}) \begin{cases} \Delta u = 0 & \text{in } \Omega \\ \frac{\partial u}{\partial \nu} = 0 & \text{on } \partial\Omega. \end{cases} \tag{4.1.40}$$

We are going to show that:

(i) (**D0**) has only the solution $u = 0$;

(ii) (**N0**) has only the solutions $u = \text{const}$.

The proof of the statements above is based on the **divergence theorem** and on the formula of integration by parts (for functions of several variables), which is one of its numerous consequences.

To motivate the use of the latter in the study of differential equations, consider the very simple one-dimensional BVP

$$\begin{cases} u'' = 0 & \text{in }]a, b[\\ u(a) = u(b) = 0, \end{cases} \tag{4.1.41}$$

whose unique solution is $u = 0$. One way to prove this fact is precisely using integration by parts, for if u solves (4.1.41), then

$$0 = \int_a^b u''(x)u(x)dx = [u'(x)u(x)]_a^b - \int_a^b u'^2(x)\,dx,$$

whence the last integral is zero, implying that $u' = 0$ in $[a, b]$ and therefore that $u = 0$ by the conditions at the endpoints.

Remarks on the assumptions on Ω

- Ω bounded $\Rightarrow \overline{\Omega}$ closed + bounded $\Rightarrow \overline{\Omega}$ **compact**.
- Ω open $\Rightarrow \overline{\Omega} = \Omega \cup \partial\Omega$ (**disjoint** union); recall that in general, $\overline{\Omega} = \Omega \cup \partial\Omega = \text{int}(\Omega) \cup \partial\Omega$, where the latter – but not necessarily the former – union is disjoint.
- Ω **is connected.**

Remarks on connectedness

Recall that a set $A \subset \mathbb{R}^n$ is said to be **disconnected** if it is the union of two relatively open, non-empty, disjoint subsets; it is said to be **connected** if it is not disconnected. Among the important properties of the connected subsets of $\mathbb{R}^n$ we list the following:

- If A is **convex** (or, more generally, **pathwise connected**), then it is connected.
- $A \subset \mathbb{R}$ is connected if and only if it is an interval.
- If $f : A \to \mathbb{R}$ is continuous and A is connected, then $f(A)$ is connected and hence an interval. It follows that f takes on **every** value z lying between **any two** of its values $f(x), f(y)$ ("intermediate value theorem").

Proofs of the above statements can be found, for instance, in Chapter 2, Section 2.6.

We add here without proof a useful statement, generalizing to functions of several variables the well-known fact that a differentiable function having derivative zero **on an interval** is constant.

Proposition 4.1.1. *If $\Omega \subset \mathbb{R}^n$ is open and connected and $f : \Omega \to \mathbb{R}$ has partial derivatives $\frac{\partial f}{\partial x_i}$ equal to zero in Ω for every i, $1 \le i \le n$, then f is constant.*

The divergence theorem (Gauss' theorem)

Theorem 4.1.1. *Let Ω be an open bounded connected subset of $\mathbb{R}^n$ $(n = 2, 3)$ with sufficiently regular boundary $\partial\Omega$ and let $\mathbf{F} \in C^1(\overline{\Omega}; \mathbb{R}^n)$. Then*

$$\int_{\overline{\Omega}} \operatorname{div} \mathbf{F} \, d\mathbf{x} = \int_{\partial\Omega} \mathbf{F} \cdot \mathbf{v} \, d\sigma, \tag{4.1.42}$$

where $\mathbf{v}$ is the outward pointing normal versor to $\partial\Omega$.

Remark. By "sufficiently regular" boundary $\partial\Omega$ we mean **for instance**:

- $(n = 2)$ $\partial\Omega$ is a simple **closed** regular curve;
- $(n = 3)$ $\partial\Omega$ is a simple **closed** regular **orientable** surface, the latter meaning that there exists a **continuous** function $\mathbf{v} : \partial\Omega \to \mathbb{R}^n$ representing the outward pointing normal versor to $\partial\Omega$.

Moreover, in the above, the word **closed** does not refer to its usual (metric/topological) sense, but to the (geometric-differential) sense of "closed curve" or "boundaryless surface," the easiest examples of these being respectively a circle $(n = 2)$ or a spherical surface $(n = 3)$.

Remark. Formula (4.1.42) relates:

- $(n = 2)$ a **double integral** with a **line integral**,
- $(n = 3)$ a **triple integral** with a **surface integral**.

We can ideally complete the above equalities with that holding for the case $n = 1$, that is, with the familiar integration formula

$$\int_a^b f'(x)\,dx = f(b) - f(a),$$

which holds for every $f \in C^1([a,b])$.

Remark. For a proof of the divergence theorem, see for instance [18] or [19].

Theorem 4.1.2. *Let Ω be an open, bounded, and connected subset of $\mathbb{R}^n$ $(n = 2,3)$ for which the divergence theorem holds. Then:*
- ***(D0)** has only the solution $u = 0$;*
- ***(N0)** has only the solutions $u = \text{const.}$*

***Consequently**, on the basis of "General Remark" **1.**,*
- ***(D)** has **at most one** solution;*
- *If **(N)** has a solution, this is not unique but is determined modulo an additive constant.*

The proof of Theorem 4.1.2 rests on the **formula of integration by parts** for functions of several variables, which is obtained as a consequence of the divergence theorem (Theorem 4.1.1).

Exercise 4.1.3. Check that

$$\operatorname{div}(v\mathbf{z}) = \nabla v \cdot \mathbf{z} + v(\operatorname{div}\mathbf{z}) \tag{4.1.43}$$

for any $v \in C^1(\Omega)$, $\mathbf{z} \in C^1(\Omega, \mathbb{R}^n)$.

Exercise 4.1.4 (Formula of integration by parts). Let $v \in C^1(\overline{\Omega})$, $\mathbf{z} \in C^1(\overline{\Omega}, \mathbb{R}^n)$. Then

$$\int_{\overline{\Omega}} \nabla v \cdot \mathbf{z}\,d\mathbf{x} = \int_{\partial\Omega} v\mathbf{z} \cdot \boldsymbol{\nu}\,d\sigma - \int_{\overline{\Omega}} v(\operatorname{div}\mathbf{z})\,d\mathbf{x}. \tag{4.1.44}$$

This follows immediately from (4.1.43), on integrating over $\overline{\Omega}$ and using the divergence theorem.

Now let $u \in C^2(\overline{\Omega})$ and take $\mathbf{z} = \nabla u$ in (4.1.44). Then we obtain the formula

$$\int_{\overline{\Omega}} \nabla v \cdot \nabla u\,d\mathbf{x} = \int_{\partial\Omega} v\frac{\partial u}{\partial \boldsymbol{\nu}}\,d\sigma - \int_{\overline{\Omega}} v(\Delta u)\,d\mathbf{x},$$

which can also be written as

$$\int_{\overline{\Omega}} (\Delta u)v\,d\mathbf{x} = \int_{\partial\Omega} \frac{\partial u}{\partial \boldsymbol{\nu}} v\,d\sigma - \int_{\overline{\Omega}} \nabla u \cdot \nabla v\,d\mathbf{x} \tag{4.1.45}$$

and holds for any pair of functions u, v with $u \in C^2$ and $v \in C^1$; it is usually called the **Green's formula**. In particular, putting $v = u$ in (4.1.45), we have

$$\int_{\overline{\Omega}} (\Delta u) u \, d\mathbf{x} = \int_{\partial\Omega} \frac{\partial u}{\partial \nu} u \, d\sigma - \int_{\overline{\Omega}} \|\nabla u\|^2 \, d\mathbf{x} \tag{4.1.46}$$

for any $u \in C^2(\overline{\Omega})$.

Proof of Theorem 4.1.2

Let u be a solution of either **(D0)** or **(N0)**. We suppose, in addition to the stated hypotheses, that $u \in C^2(\overline{\Omega})$.

From (4.1.46), necessarily $\int_{\overline{\Omega}} \|\nabla u\|^2 \, d\mathbf{x} = 0$. By the continuity of ∇u and the n-dimensional version of Proposition 3.1.1, it thus follows that

$$\nabla u(\mathbf{x}) = 0 \quad \forall \mathbf{x} \in \overline{\Omega}.$$

As Ω is **connected** by assumption, it first follows that

$$u(\mathbf{x}) = \text{const} \equiv C \quad \forall \mathbf{x} \in \Omega$$

and then, by continuity, that

$$u(\mathbf{x}) = C \quad \forall \mathbf{x} \in \overline{\Omega}.$$

Thus, we conclude that any solution u of either **(D0)** or **(N0)** is necessarily constant in $\overline{\Omega}$. If in particular u is a solution of **(D0)**, so that $u = 0$ on $\partial\Omega$, it follows that $u = 0$ on $\overline{\Omega}$.

4.2 The Dirichlet problem for the Laplace equation in a rectangle (I)

Consider the problem

$$(\mathbf{D}) \begin{cases} \Delta u = f & \text{in } \Omega \\ u = g & \text{on } \partial\Omega \end{cases} \tag{4.2.1}$$

where Ω is an open bounded connected subset of $\mathbb{R}^n (n = 2, 3)$ with sufficiently regular boundary $\partial\Omega$, while $f : \Omega \to \mathbb{R}$ and $g : \partial\Omega \to \mathbb{R}$ are given continuous functions: these are the **data** of the problem (**D**). A **solution** of (**D**) is a function $u \in C^2(\Omega) \cap C(\overline{\Omega})$ satisfying pointwisely both the equation in Ω and the boundary condition on $\partial\Omega$.

As we have seen, problem (D) has **at most one solution**.

Question. Does there actually exist a solution of (D)?

This is not a simple problem to solve in general: the (affirmative) answer requires advanced methods of functional analysis, in particular the study of **Sobolev spaces**; see, for instance, [14].

One can prove the existence of a solution (and in fact construct it explicitly) by elementary methods in some **particular situation**:
- $f = 0$ (that is, the homogeneous equation, named the **Laplace equation**);
- $n = 2$;
- Ω is a **rectangle** or a **circle**.

Methods: separation of variables + Fourier series

Consider the case

$$\Omega = \text{an open rectangle } R =]a, b[\times]c, d[$$

so that

$$\overline{\Omega} = \text{the closed rectangle } \overline{R} = [a, b] \times [c, d],$$

$$\partial\Omega = \partial R \equiv K = \bigcup_{i=1}^{4} K_i.$$

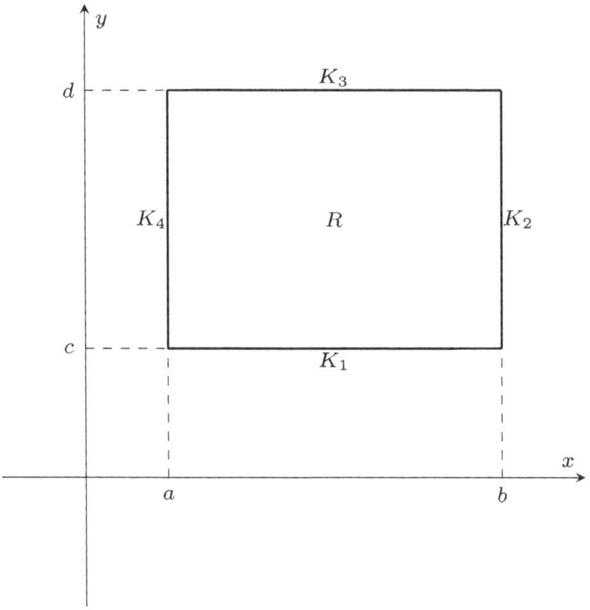

Summarizing, our problem is

$$(\mathbf{D})\begin{cases}\Delta u = 0 & \text{in } R\\ u = g & \text{on } K\end{cases}\tag{4.2.2}$$

and we look for a solution $u \in C^2(R) \cap C(\overline{R})$.

On the basis of the "superposition principle" ("General Remark" **2.**) it will be enough to solve simpler problems such as

$$(\mathbf{D1})\begin{cases}\Delta u = 0 & \text{in } R\\ u = g & \text{on } K_1\\ u = 0 & \text{on } K_2 \cup K_3 \cup K_4.\end{cases}\tag{4.2.3}$$

Indeed, suppose that u_1 is a solution of **(D1)** and let u_2, u_3, u_4 be solutions of, respectively, the problems

$$(\mathbf{D2})\begin{cases}\Delta u = 0 & \text{in } R\\ u = g & \text{on } K_2\\ u = 0 & \text{on } K \setminus K_2,\end{cases}\qquad (\mathbf{D3})\begin{cases}\Delta u = 0 & \text{in } R\\ u = g & \text{on } K_3\\ u = 0 & \text{on } K \setminus K_3,\end{cases}\qquad (\mathbf{D4})\begin{cases}\Delta u = 0 & \text{in } R\\ u = g & \text{on } K_4\\ u = 0 & \text{on } K \setminus K_4.\end{cases}$$

Then $u = u_1 + u_2 + u_3 + u_4$ is (**the**) solution of (**D**); indeed,

$$\begin{cases}\Delta u(\mathbf{x}) = 0 & \forall \mathbf{x} \in R\\ u(\mathbf{x}) = u_1(\mathbf{x}) + u_2(\mathbf{x}) + u_3(\mathbf{x}) + u_4(\mathbf{x}) = g(\mathbf{x}) & \forall \mathbf{x} \in K,\end{cases}$$

the second equality holding because every $\mathbf{x} \in K$ belongs to precisely one K_i; in the corner points $\mathbf{x}$ of K we require that $g(\mathbf{x}) = 0$, so as to have continuous boundary data in the subproblems **(D1)**–**(D4)**.

We consider thus **(D1)** and suppose for convenience that

$$R =]0, \pi[\times]0, 1[$$

so that our problem becomes explicitly (writing f rather than g)

$$(\mathbf{D})\begin{cases}\Delta u = 0 & \text{in } R & (\mathbf{E})\\ u(x,0) = f(x) & 0 \le x \le \pi & (\mathbf{C}_1)\\ u(\pi,y) = u(x,1) = u(0,y) = 0. & & (\mathbf{C}_2)\end{cases}\tag{4.2.4}$$

Let us consider first the differential equation **(E)**. We seek solutions u of the form

$$u(x,y) = X(x)Y(y)$$

(**separation of variables**), so that assuming in addition that $u(x,y) \neq 0 \quad \forall (x,y) \in R$ we have

$$\Delta u = 0 \Leftrightarrow X''Y + XY'' = 0 \Leftrightarrow \frac{X''}{X} + \frac{Y''}{Y} = 0$$
$$\Leftrightarrow \frac{X''}{X} = \text{const.} = -\lambda, \quad \frac{Y''}{Y} = \lambda.$$

Therefore, under separation of variables (**E**) is equivalent to the system of ODEs

$$\begin{cases} X'' + \lambda X = 0 & 0 < x < \pi \\ Y'' - \lambda Y = 0 & 0 < y < 1. \end{cases}$$

These are **coupled** equations, for λ is the same in the two equations.

Boundary conditions

We have

$$(\mathbf{C}_1) \quad X(x)Y(0) = f(x) \quad 0 \le x \le \pi,$$
$$(\mathbf{C}_2) \quad \begin{cases} X(\pi)Y(y) = 0 & 0 \le y \le 1 \\ X(x)Y(1) = 0 & 0 \le x \le \pi \\ X(0)Y(y) = 0 & 0 \le y \le 1. \end{cases}$$

($\mathbf{C}_2$) is satisfied requiring that $X(\pi) = X(0) = 0$ and that $Y(1) = 0$.

Summing up, we see that (**E**) and ($\mathbf{C}_2$) are satisfied by $u(x,y) = X(x)Y(y)$ if X and Y satisfy respectively the problems

$$(\mathbf{A}) \begin{cases} X'' + \lambda X = 0 & 0 < x < \pi \\ X(0) = X(\pi) = 0, \end{cases} \qquad (\mathbf{B}) \begin{cases} Y'' - \lambda Y = 0 & 0 < y < 1 \\ Y(1) = 0. \end{cases}$$

Question. Do there exist **non-zero** solutions of **A** and **B**?

It is shown in Exercise 3.5.1 that
(**A**) has solutions $\neq 0$ if and only if $\lambda = \lambda_n = n^2$, and the corresponding solutions are

$$X_n(x) = C \sin nx.$$

Moreover, we will soon check (Exercise 4.2.1) that for $\lambda = \lambda_n$, the solutions of (**B**) are

$$Y_n(x) = C \sinh n(1 - y).$$

Conclusion

For every $n \in \mathbb{N}$, the function

$$u_n(x,y) = \sin nx \sinh n(1 - y)$$

is a solution of (**E**) + ($\mathbf{C}_2$).

It follows that any finite sum $s_N = \sum_{n=1}^{N} c_n u_n$ has the same property.

We look for a solution of the form

$$u(x,y) = \sum_{n=1}^{\infty} c_n u_n(x,y) = \sum_{n=1}^{\infty} c_n \sin nx \sinh n(1-y) \tag{4.2.5}$$

(see "General Remark" **3.**), where the constants $c_n (n \in \mathbb{N})$ have to be chosen in such a way that:

- the series (4.2.5) is convergent at every point of $(x,y) \in \overline{R}$;
- the condition ($\mathbf{C}_1$) is satisfied.

After that, it will be necessary to verify that the sum function u is actually a solution of (**D**), that is,

(i) $u \in C(\overline{R})$ and satisfies ($\mathbf{C}_1$) and ($\mathbf{C}_2$);
(ii) $u \in C^2(R)$ and satisfies (**E**).

Let us impose formally on u that it satisfies ($\mathbf{C}_1$):

$$u(x,0) = \sum_{n=1}^{\infty} c_n \sin nx \sinh n = f(x), \quad 0 \le x \le \pi. \tag{4.2.6}$$

Definition 4.2.1. Given $f : [0,\pi] \to \mathbb{R}$, f piecewise continuous in $[0,\pi]$, the series

$$\sum_{n=1}^{\infty} b_n \sin nx, \quad b_n = \frac{2}{\pi} \int_0^{\pi} f(x) \sin nx \, dx \quad (n \in \mathbb{N}) \tag{4.2.7}$$

is called the **sine Fourier series of** f.

We shall choose the coefficients c_n so that the series in (4.2.6) is the sine Fourier series of f:

$$c_n \sinh n = b_n = \frac{2}{\pi} \int_0^{\pi} f(x) \sin nx \, dx \quad (n \in \mathbb{N}). \tag{4.2.8}$$

The condition (4.2.6) (that is, **C1**) means that this series must converge to f pointwisely on $[0,\pi]$.

Remark 4.2.1. The Fourier sine series of an $f : [0,\pi] \to \mathbb{R}$ is nothing but the Fourier series of the **odd** extension f_d of f to $[-\pi,\pi]$, further extended by 2π-periodicity to all of $\mathbb{R}$ (see Figure 4.1). Consistently with the notations formerly employed, we shall denote with $\hat{f}_d$ such extension of f.

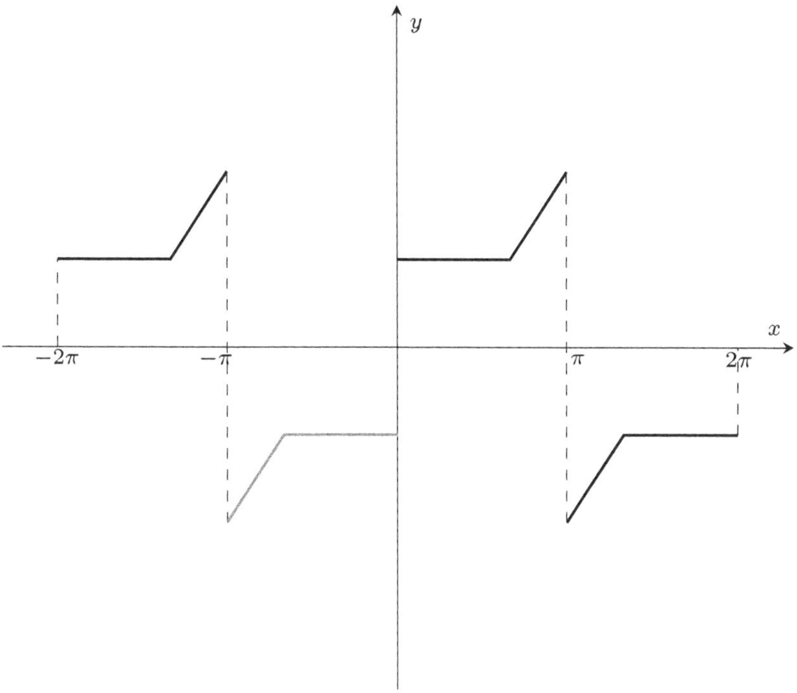

Figure 4.1: Extending f to f_d and then to $\hat{f}_d$.

Exercise 4.2.1. Consider the problem

$$(\mathbf{B})\begin{cases} Y'' - \lambda Y = 0 & (0 < y < 1) \\ Y(1) = 0. \end{cases}$$

We are interested in solving (**B**) for $\lambda > 0$. Putting $\beta = \sqrt{\lambda}$, the general solution of the differential equation in (**B**) is

$$Y(y) = Ce^{\beta y} + De^{-\beta y} \quad (C, D \in \mathbb{R}).$$

Imposing the condition at $y = 1$, we have

$$Y(1) = Ce^{\beta} + De^{-\beta} = 0 \Longrightarrow D = -Ce^{2\beta},$$

so that

$$\begin{aligned} Y(y) &= C[e^{\beta y} - e^{2\beta}e^{-\beta y}] \\ &= Ce^{\beta}[e^{-\beta}e^{\beta y} - e^{\beta}e^{-\beta y}] \\ &= 2Ce^{\beta}\frac{[e^{-\beta(1-y)} - e^{\beta(1-y)}]}{2} \\ &= C' \sinh\beta(1-y) \quad (C' = -2Ce^{\beta}), \end{aligned}$$

where the **hyperbolic sine** of x, $\sinh x$, is defined by the equation

$$\sinh x \equiv \frac{e^x - e^{-x}}{2} \qquad (x \in \mathbb{R}).$$

Thus, for $\lambda = \lambda_n = n^2$, the solutions of (**B**) are

$$Y_n(y) = C \sinh n(1-y).$$

4.3 Pointwise and uniform convergence of Fourier series

In order to guarantee the pointwise and unifom convergence of the Fourier series of a 2π-periodic function f, the assumption that $f \in \hat{C}_{2\pi}(\mathbb{R})$, or even that $f \in C_{2\pi}(\mathbb{R})$, is not enough; some form of **differentiability** is required.

In this section, we state two theorems regarding these questions. For the first (regarding pointwise convergence) we give no proof but just some comments and examples; the second one, concerning uniform convergence, will instead be proved as it employs in a nice way much of the information previously acquired.

Theorem 4.3.1. *Let $f \in \hat{C}_{2\pi}(\mathbb{R})$ and suppose that at the point $x_0 \in \mathbb{R}$ there exist and are finite the limits*

$$f'(x_0^-) \equiv \lim_{x \to x_0^-} \frac{f(x) - f(x_0^-)}{x - x_0}, \quad f'(x_0^+) \equiv \lim_{x \to x_0^+} \frac{f(x) - f(x_0^+)}{x - x_0}, \tag{4.3.1}$$

*called respectively the **left- and right-pseudoderivative** of f at x_0. Then the Fourier series of f evaluated at the point x_0 converges to $f(x_0)$.*

For a proof of Theorem 4.3.1 see, e. g., [9]. Figure 4.2 shows an f that does not satisfy the assumptions of this theorem at the point $x_0 = 0$ – even if we regularize it putting $f(0) = 1/2$.

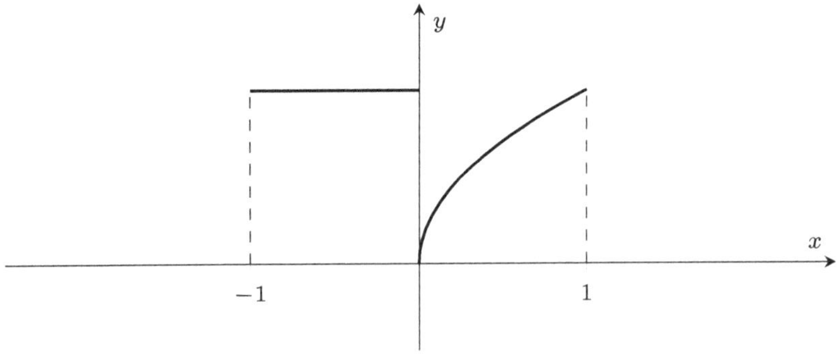

Figure 4.2: $f(x) = 1\ (-1 \le x < 0), = \sqrt{x}\ (0 < x \le 1)$.

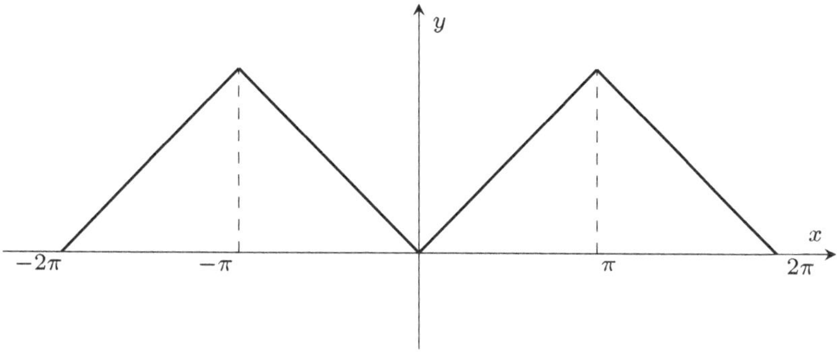

Figure 4.3: $f(x) = |x|, -\pi \leq x \leq \pi$.

Remark 4.3.1. If f is continuous at x_0, the limits in (4.3.1) (when they exist) are the *left- and right-derivative* of f at x_0. In particular, f is differentiable at $x_0 \Longrightarrow f$ is continuous at x_0 and $f'(x_0^-) = f'(x_0^+) = f'(x_0) \Longrightarrow$ the Fourier series of f converges in x_0 to $f(x_0)$.

Remark 4.3.2. If the assumptions of Theorem 4.3.1 hold except that f is not regular at x_0, then the Fourier series of f converges in x_0 to the mean value

$$\frac{f(x_0^+) + f(x_0^-)}{2}.$$

Remark 4.3.3. If f is differentiable in $]x_0 - \delta, x_0[$ $(\delta > 0)$ and there exists the limit $\lim_{x \to x_0^-} f'(x)$, then $f'(x_0^-)$ exists too, and they are equal; this follows at once applying the L'Hopital rule. (Note that we would usually define $f'(x_0^-)$ and $f'(x_0^+)$ precisely via these limits, rather than through the formulae in (4.3.1).)

Examples with numerical applications

Exercise 4.3.1. In the light of Theorem 4.3.1 and of the above remarks, discuss the pointwise convergence of the Fourier series of the functions considered in Examples 3.2.4 and 3.2.5.

Example 4.3.1. Consider $\hat{f}$ when f is the absolute value function for $|x| \leq \pi$ (see Figure 4.3).

For this $\hat{f}$ we have seen in Example 3.2.6 that $b_k = 0 \quad \forall k \in \mathbb{N}$, while

$$a_0 = \frac{\pi}{2}, \quad a_k = -\frac{2}{k^2\pi}\left[1 - (-1)^k\right] = \begin{cases} 0 & (k \text{ even}) \\ -\frac{4}{k^2\pi} & (k \text{ odd}). \end{cases}$$

Therefore, using Theorem 4.3.1 at $x_0 = 0$, we obtain

$$0 = \frac{\pi}{2} - \sum_{k=1}^{\infty} \frac{4}{(2k+1)^2\pi},$$

whence

$$\sum_{k=1}^{\infty} \frac{1}{(2k+1)^2} = \frac{\pi^2}{8}.$$

Suppose now that the hypotheses of Theorem 4.3.1 are satisfied at each point of $\mathbb{R}$, so that we have pointwise convergence of the Fourier series of f to f itself. What further condition must we impose on f in order to have **uniform** convergence?

Remark. The properties of uniform convergence imply that if f is not continuous at just one point x_0, then the Fourier series of f cannot converge uniformly to f. In other words, a **necessary** condition for uniform convergence is that $f \in C_{2\pi}(\mathbb{R})$.

Remark. For periodic functions, the uniform convergence on $\mathbb{R}$ is the same as uniform convergence on $[0, 2\pi]$ (or any interval of length 2π); indeed, for such functions,

$$\sup_{x\in\mathbb{R}} |f(x)| = \sup_{x\in[0,2\pi]} |f(x)|.$$

Piecewise C^1 functions

Definition 4.3.1. A function $f : [a, b] \to \mathbb{R}$ is said to be **piecewise of class** C^1 (briefly, piecewise C^1) in $[a, b]$ if:
(i) f is continuous in $[a, b]$;
(ii) f is differentiable in $[a, b]$ except at most in a finite number of points $x_1, \ldots, x_n$ of $[a, b]$;
(iii) the derivative f' of f is continuous in $[a, b] \setminus \{x_1, \ldots, x_n\}$ and the limits

$$\lim_{x\to x_i^\pm} f'(x) \quad (i = 1, \ldots, n) \tag{4.3.2}$$

exist and are finite.

For instance, the function shown in Figure 4.4 is *not* piecewise C^1, while the function $f(x) = |x|$ is. More generally, the "broken lines" (see Figure 4.5) are graphs of piecewise

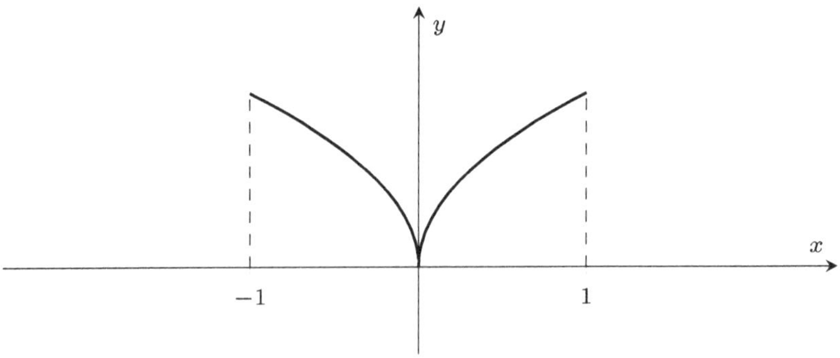

Figure 4.4: $\sqrt{|x|}$.

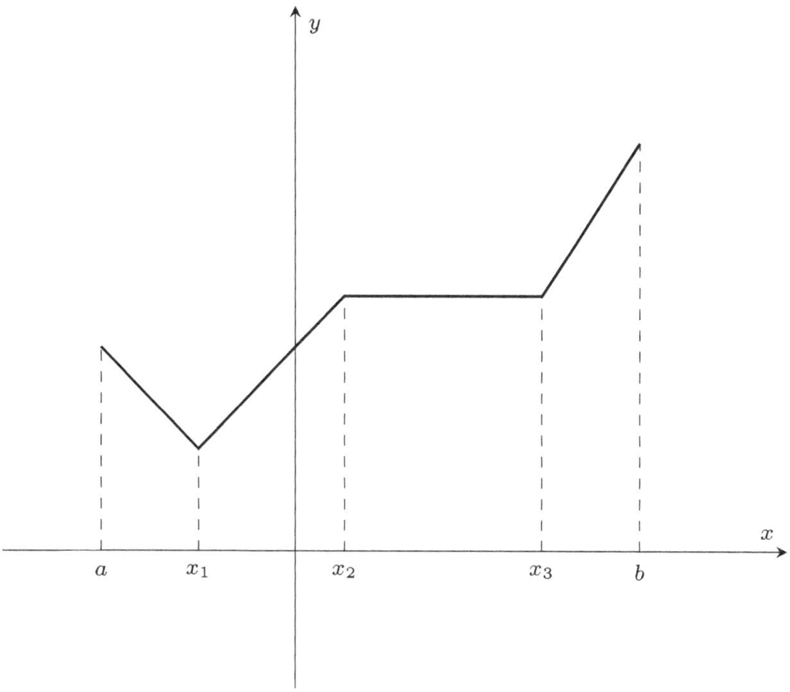

Figure 4.5: A broken line.

C^1 functions. In more precise terms, continuous, piecewise linear functions are (basic) examples of piecewise C^1 functions.

Remark 4.3.4. In Definition 4.3.1, some authors replace the requirement (i) of continuity with that of piecewise continuity.

Proposition 4.3.1. *Let $f \in \hat{C}_{2\pi}(\mathbb{R})$. If f is piecewise C^1, then its Fourier series converges pointwisely to f on $\mathbb{R}$.*

Proof. Indeed, if f is piecewise C^1, then by definition for every $x_0 \in \mathbb{R}$ the limits

$$\lim_{x\to x_0^-} f'(x), \quad \lim_{x\to x_0^+} f'(x)$$

exist and are finite. Therefore, on the basis of Remark 4.3.3, it follows that also the limits $f'(x_0^-), f'(x_0^+)$ defined in (4.3.1) exist and are equal to the previous ones. Thus, by Theorem 4.3.1, the Fourier series of f in x_0 converges to $f(x_0)$. □

Remark 4.3.5. Let $f \in \hat{C}_{2\pi}(\mathbb{R})$. If f is piecewise C^1, then its derivative f' can be **defined** also at the finitely many points $x_1, \dots, x_n \in [0, 2\pi]$ in which f is not differentiable, by assigning to $f'(x_i)$ the mean value of the limits in (4.3.2) and thus obtaining a regular f'. Moreover, the f' thus obtained is also 2π-periodic, for

$$f'(x_0+2\pi) = \lim_{h\to 0} \frac{f(x_0+2\pi+h) - f(x_0+2\pi)}{h} = f'(x_0).$$

In conclusion, $f' \in \hat{C}_{2\pi}(\mathbb{R})$, so we can consider the Fourier series of f'.

Lemma 4.3.1. *Let $f \in \hat{C}_{2\pi}(\mathbb{R})$ and suppose in addition that f is piecewise C^1. Let $a_0, a_k, b_k (k \in \mathbb{N})$ be the Fourier coefficients of f and let $\alpha_0, \alpha_k, \beta_k (k \in \mathbb{N})$ be the Fourier coefficients of f':*

$$\alpha_0 = \frac{1}{2\pi}\int_0^{2\pi} f'(x)\,dx, \quad \alpha_k = \frac{1}{\pi}\int_0^{2\pi} f'(x)\cos kx\,dx, \quad \beta_k = \frac{1}{\pi}\int_0^{2\pi} f'(x)\sin kx\,dx. \tag{4.3.3}$$

Then

$$\begin{cases} \alpha_0 = 0 \\ \alpha_k = kb_k \\ \beta_k = -ka_k. \end{cases} \tag{4.3.4}$$

Proof. This is left as an exercise. □

Remark 4.3.6. The rule of integration by parts for two functions f, g holds not only if $f, g \in C^1[a,b]$, but also, more generally, if f, g are piecewise C^1 in $[a,b]$.

Theorem 4.3.2. *Let $f \in \hat{C}_{2\pi}(\mathbb{R})$. If in addition f is piecewise C^1, then its Fourier series converges **uniformly** to f on $\mathbb{R}$.*

Proof. On the basis of Proposition 4.3.1, we know already that the Fourier series of f converges **pointwisely** to f on $\mathbb{R}$. To prove that the convergence is uniform, we apply to the Fourier series of f the "Weierstrass M-test" (see Proposition 2.1.1), stating that given a sequence (f_n) of bounded functions defined on a set A, if there exists a sequence (M_n), with $M_n \geq 0$ for all $n \in \mathbb{N}$, such that:

(i) $|f_n(x)| \leq M_n$ for all $x \in A$ and all $n \in \mathbb{N}$ and
(ii) the series $\sum_{n=1}^{\infty} M_n$ is convergent,

then the series $\sum_{n=1}^{\infty} f_n$ converges uniformly in A.

The Fourier series of f is $\sum_{k=0}^{\infty} f_k$ with $f_k(x) = a_k \cos kx + b_k \sin kx$. Using Lemma 4.3.1 we have for every $x \in \mathbb{R}$ and every $k \in \mathbb{N}$

$$|f_k(x)| = |a_k \cos kx + b_k \sin kx| \leq |a_k| + |b_k| = \frac{|\beta_k|}{k} + \frac{|\alpha_k|}{k},$$

whence, recalling the inequality $2cd \leq c^2 + d^2$, we obtain

$$|f_k(x)| \leq \frac{1}{2}\left(\frac{1}{k^2} + \beta_k^2\right) + \frac{1}{2}\left(\frac{1}{k^2} + \alpha_k^2\right) = \frac{1}{k^2} + \frac{1}{2}(\alpha_k^2 + \beta_k^2) \equiv M_k.$$

Now the conclusion follows on recalling that, by Parseval's identity applied to f', the series

$$\sum_{k=1}^{\infty}(\alpha_k^2+\beta_k^2)$$

is convergent. □

Exercise 4.3.2. Let $f(x) = x^2$ ($|x| \le \pi$) and let $\hat{f}$ be its 2π-periodic extension to $\mathbb{R}$ (see Figure 4.6). Prove that

$$\hat{f}(x) = \frac{\pi^2}{3} + 4\sum_{k=1}^{\infty}\frac{(-1)^k}{k^2}\cos kx \quad (x \in \mathbb{R}). \tag{4.3.5}$$

In particular, taking $x = 0$ and then $x = \pi$ in (4.3.5), one obtains

$$\sum_{k=1}^{\infty}\frac{(-1)^k}{k^2} = -\frac{\pi^2}{12}, \quad \sum_{k=1}^{\infty}\frac{1}{k^2} = \frac{\pi^2}{6}.$$

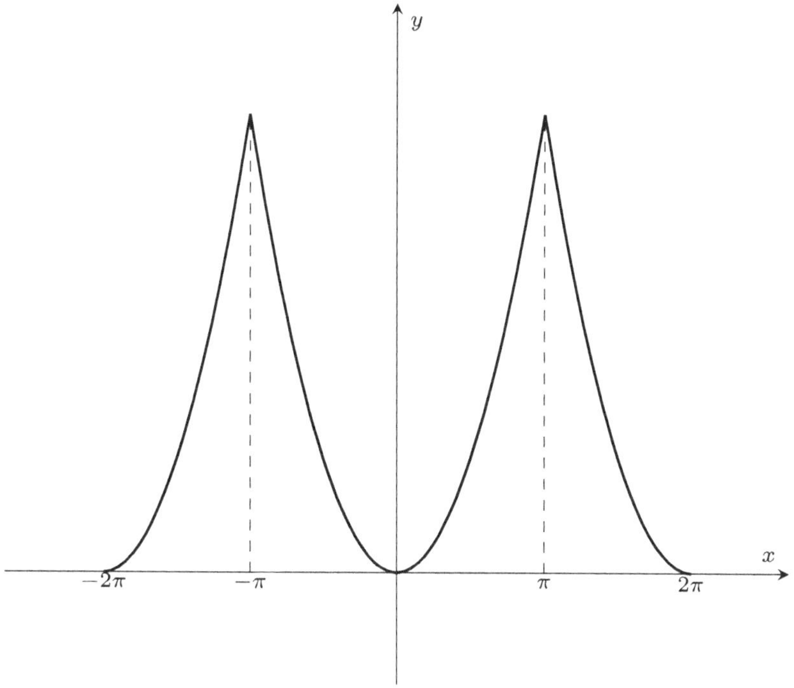

Figure 4.6: $f(x) = x^2$, $|x| \le \pi$.

4.4 The Dirichlet problem for the Laplace equation in a rectangle (II)

In order to complete our discussion of problem (**D**), it is useful to gain some more information on Fourier series, which should be added to Theorem 4.3.2.

Sine and cosine Fourier series

Let $f : [0,\pi] \to \mathbb{R}$ be piecewise continuous. Then f can be expanded into a **cosine Fourier series** and/or into a **sine Fourier series**. In fact, f can be continued to $[-\pi,\pi]$ as an **even** function f_p or as an **odd** function f_d as follows:

$$f_p(x) = \begin{cases} f(x) & 0 \le x \le \pi \\ f(-x) & -\pi \le x < 0, \end{cases} \qquad f_d(x) = \begin{cases} f(x) & 0 \le x \le \pi \\ -f(-x) & -\pi \le x < 0. \end{cases} \tag{4.4.1}$$

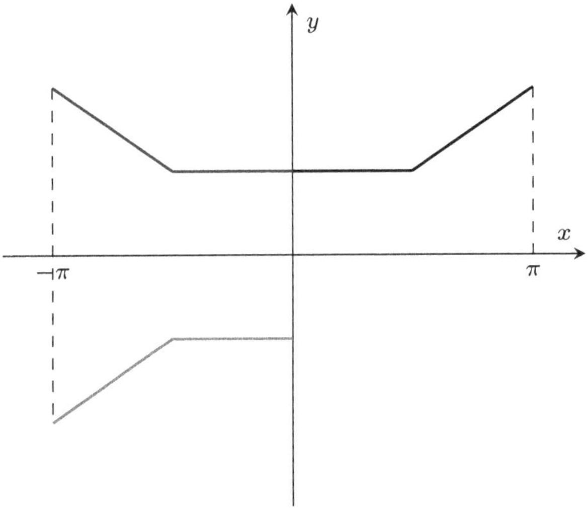

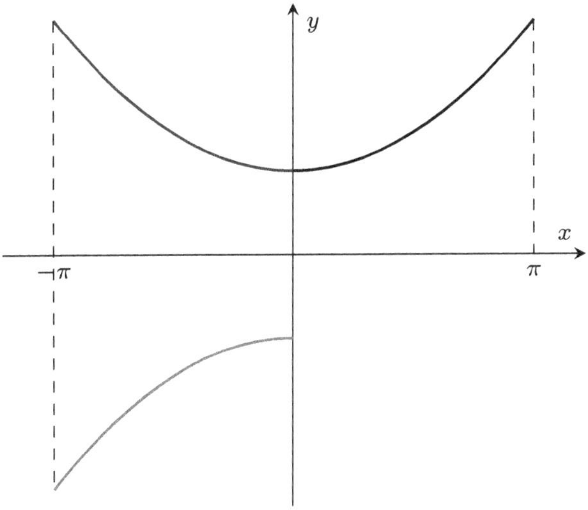

Remark. f piecewise continuous on $[0,\pi] \Longrightarrow f_p, f_d$ piecewise continuous on $[-\pi,\pi] \Longrightarrow \hat{f}_p, \hat{f}_d$ piecewise continuous on $\mathbb{R}$.

We can thus consider the Fourier series of $\hat{f}_p, \hat{f}_d$.

Definition 4.4.1. Let $f : [0,\pi] \to \mathbb{R}$ be piecewise continuous. The Fourier series of $\hat{f}_p, \hat{f}_d$ are called respectively the **cosine Fourier series** and the **sine Fourier series** of f.

With respect to the **continuity** of $\hat{f}_p, \hat{f}_d$, observe that

$$f \quad \text{continuous on} \quad [0,\pi] \Longrightarrow \begin{cases} f_p & \text{continuous on } [-\pi,\pi] \\ f_d & \text{continuous on } [-\pi,\pi] \Leftrightarrow f(0) = 0. \end{cases} \tag{4.4.2}$$

It follows in particular that

$$\begin{cases} f \text{ piecewise } C^1 \text{ on } [0,\pi] \\ f(0) = 0 \end{cases} \Longrightarrow f_d \quad \text{piecewise } C^1 \text{ on } [-\pi,\pi]$$
$$\Longrightarrow \hat{f}_d \quad \text{piecewise } C^1 \text{ on } \mathbb{R} \quad \text{iff} \quad f_d(-\pi) = f_d(\pi).$$

However, the last equality means $-f(\pi) = f(\pi)$, that is, $f(\pi) = 0$.
Summarizing, we have the following statement.

Proposition 4.4.1. *Let $f : [0,\pi] \to \mathbb{R}$ be piecewise C^1. If $f(0) = f(\pi) = 0$, then its odd and 2π-periodic extension to the whole of $\mathbb{R}$, $\hat{f}_d$, is piecewise C^1.*

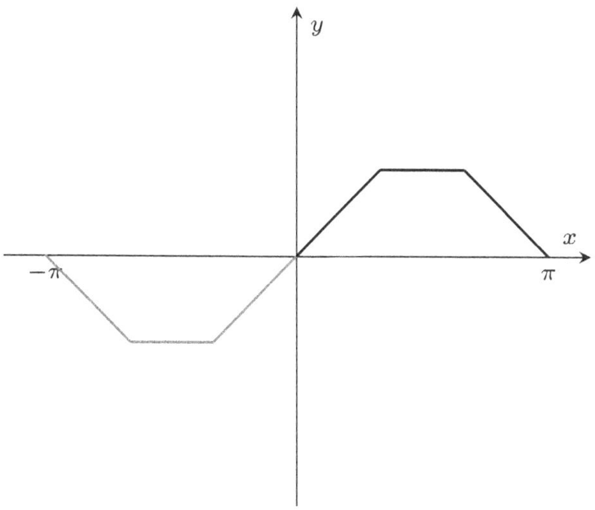

Let us introduce a useful notation. Given any interval $[a,b] \subset \mathbb{R}$, we put

$$C_0^1[a,b] = \{f \in C^1[a,b] : f(a) = f(b) = 0\}.$$

As a special case of Proposition 4.4.1, we then get the following statement.

Lemma 4.4.1. *Let $f \in C_0^1[0,\pi]$. Then its odd and 2π-periodic extension to the whole of $\mathbb{R}$, $\hat{f}_d$, is piecewise C^1.*

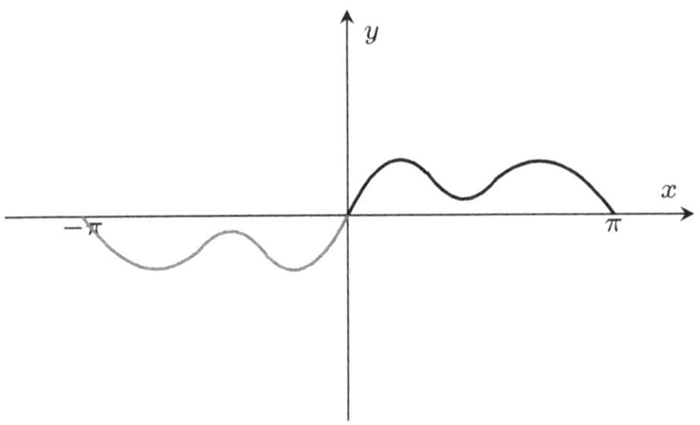

Thus, if we consider functions in $C_0^1[0,\pi]$, the previous discussion on their sine Fourier series and the uniform convergence theorem (Theorem 4.3.2) yield the following result.

Proposition 4.4.2. *Let $f \in C_0^1[0,\pi]$ and let $b_n = \frac{2}{\pi}\int_0^\pi f(x)\sin nx\,dx$ ($n \in \mathbb{N}$). Then we have*

$$\sum_{n=1}^{\infty} b_n \sin nx = f(x) \quad \textit{for every } x \in [0,\pi]. \tag{4.4.3}$$

Moreover, the convergence of the series in (4.4.3) *is **uniform** in* $[0,\pi]$.

Let us return to the series in the two variables x, y presented in equation (4.2.5),

$$\sum_{n=1}^{\infty} c_n u_n(x,y) = \sum_{n=1}^{\infty} c_n \sin nx \sinh n(1-y), \tag{4.4.4}$$

as a candidate solution for our problem (**D**). We now know that taking $f \in C_0^1[0,\pi]$ in (**D**) and taking c_n determined by the condition

$$c_n \sinh n = b_n = \frac{2}{\pi}\int_0^\pi f(x)\sin nx\,dx \quad (n \in \mathbb{N}), \tag{4.4.5}$$

that is, considering the series

$$\sum_{n=1}^{\infty} b_n \sin nx \frac{\sinh n(1-y)}{\sinh n}, \tag{4.4.6}$$

this converges (to f) on the "lower side" K_1 of the closed rectangle $\overline{R}$.

Problem. Do we have convergence in $\overline{R}$?

Proposition 4.4.3. *Let $f \in C_0^1[0,\pi]$ and let $b_n = \frac{2}{\pi}\int_0^{\pi} f(x)\sin nx\,dx$ $(n \in \mathbb{N})$. Then we have*

$$\sum_{n=1}^{\infty} |b_n| < \infty. \tag{4.4.7}$$

This follows from Lemma 4.4.1 and from a more careful use of Theorem 4.3.2. Indeed, in that theorem we have proved that if $h \in \hat{C}_{2\pi}(\mathbb{R})$ is piecewise of class C^1, then its Fourier coefficients a_n, b_n satisfy the condition

$$\sum_{n=1}^{\infty} (|a_n| + |b_n|) < \infty, \tag{4.4.8}$$

as a consequence of the Parseval identity. Now if we start from a function $f \in C_0^1[0,\pi]$, by Lemma 4.4.1 its odd 2π-periodic extension $\hat{f}_d$ is piecewise C^1; therefore, its Fourier coefficients will enjoy the property (4.4.8), which in fact reduces to (4.4.7) because $a_n = 0$ by virtue of the oddness of $\hat{f}_d$.

We can now answer the convergence question positively. We have indeed the following corollary.

Corollary 4.4.1. *Let $f \in C_0^1[0,\pi]$. Then the series*

$$\sum_{n=1}^{\infty} b_n \sin nx \frac{\sinh n(1-y)}{\sinh n}, \qquad b_n = \frac{2}{\pi}\int_0^{\pi} f(x)\sin nx\,dx, \tag{4.4.9}$$

converges ***uniformly*** *in $[0,\pi]\times[0,1] = \overline{R}$.*

Proof. This follows from Proposition 4.4.3 and from the Weierstrass M-test (in two variables); indeed, we have

$$\left| b_n \sin nx \frac{\sinh n(1-y)}{\sinh n} \right| \le K|b_n| \quad \forall (x,y) \in \overline{R} \tag{4.4.10}$$

for all $n \in \mathbb{N}$ and for some $K > 0$, because we have

$$0 \le \frac{\sinh n(1-y)}{\sinh n} \le Ke^{-ny} \le K \tag{4.4.11}$$

as follows from the relations (holding for $0 \le y \le 1$)

$$\frac{\sinh n(1-y)}{\sinh n} = \frac{e^{n(1-y)} - e^{-n(1-y)}}{e^n - e^{-n}}$$
$$= \frac{e^{n(1-y)}[1-e^{-2n(1-y)}]}{e^n[1-e^{-2n}]} = e^{-ny}\frac{[.]}{[.]} \le \frac{e^{-ny}}{1-e^{-2}} \equiv Ke^{-ny}.$$

It is now time to collect the results obtained so far about problem (**D**), and also some still to be proved, in a comprehensive statement. □

Theorem 4.4.1. *Let $f \in C_0^1[0,\pi]$ and let u be the function defined in $\overline{R} = [0,\pi] \times [0,1]$ putting*

$$u(x,y) = \sum_{n=1}^{\infty} b_n \sin nx \frac{\sinh n(1-y)}{\sinh n}, \quad (x,y) \in \overline{R}, \tag{4.4.12}$$

where $b_n = \frac{2}{\pi}\int_0^\pi f(x)\sin nx\,dx$. Then u is (the) solution of (**D**).

Proof. (i) u is continuous in $\overline{R}$ by Corollary 4.4.1 and by the properties of uniform convergence. Moreover, u satisfies by construction the boundary conditions ($\mathbf{C}_1$), ($\mathbf{C}_2$); for instance,

$$u(\pi,y) = \sum_{n=1}^{\infty} b_n \sin n\pi \frac{\sinh n(1-y)}{\sinh n} = 0, \quad 0 \le y \le 1,$$

and so forth.

(ii) u is of class C^2 in the open rectangle R and satisfies the differential equation (**E**); indeed,

$$u = \sum_{n=1}^{\infty} b_n v_n \quad \text{with } v_n \in C^2(R) \quad \text{and} \quad \Delta v_n = 0 \quad \forall n \in \mathbb{N}$$

so that we will have

$$\Delta u = \Delta\left(\sum_{n=1}^{\infty} b_n v_n\right) = \sum_{n=1}^{\infty} b_n \Delta v_n = 0$$

provided we check that u has actually partial derivatives up to the second order and that the intermediate equality is true. □

Following H. Weinberger's book [2], we can say that "the series for

$$\frac{\partial u}{\partial x}, \frac{\partial u}{\partial y}, \frac{\partial^2 u}{\partial x^2}, \frac{\partial^2 u}{\partial y^2}, \frac{\partial^2 u}{\partial x \partial y}$$

are all dominated by $\sum_{n=1}^{\infty} n^2 e^{-ny}$ and therefore converge uniformly for $y \ge y_0$, whatever is $y_0 > 0$. It follows that these derivatives of u exist in R and may be obtained by term-by term differentiation of the series (4.4.12). Since each term of the series satisfies the Laplace equation, the same is true for u."

Rather than proving the above statement in all details, we illustrate the underlying main idea using the remarks that follow.

1. Regularity of functions defined as the sum of a series

The basic result on "term-by-term differentiation" of a series of functions is Corollary 2.1.3, stated and proved in the section dealing with uniform convergence and its consequences (Chapter 2, Section 2.1).

The corresponding result in two variables reads as follows.

Theorem 4.4.2. *Let $(f_n) \subset C^1(A)$, A an open subset of $\mathbb{R}^2$. Suppose that:*
- *$\sum_{n=1}^{\infty} f_n$ converges uniformly in A; let $f(x,y) = \sum_{n=1}^{\infty} f_n(x,y)$;*
- *the two series $\sum_{n=1}^{\infty} \frac{\partial f_n}{\partial x}, \sum_{n=1}^{\infty} \frac{\partial f_n}{\partial y}$ converge uniformly in A.*

Then $f \in C^1(A)$ and we have

$$\begin{cases} \frac{\partial f}{\partial x} = \sum_{n=1}^{\infty} \frac{\partial f_n}{\partial x} \\ \frac{\partial f}{\partial y} = \sum_{n=1}^{\infty} \frac{\partial f_n}{\partial y} \end{cases} \quad \text{in } A. \tag{4.4.13}$$

2. Specific bounds for the derivatives in problem (D)

In our case we have

$$u = \sum_{n=1}^{\infty} f_n = \sum_{n=1}^{\infty} b_n v_n, \quad v_n(x,y) = \sin nx \frac{\sinh n(1-y)}{\sinh n}.$$

Hence,

$$\frac{\partial u}{\partial x} : \quad \frac{\partial f_n}{\partial x}(x,y) = b_n n \cos nx \frac{\sinh n(1-y)}{\sinh n}$$

so that, putting $\beta = \frac{2}{\pi} \int_0^{2\pi} |f(x)|\, dx$ and using (4.4.11), we have

$$\left| \frac{\partial f_n}{\partial x}(x,y) \right| \le |b_n| n K e^{-ny} \le \beta n K e^{-ny} \equiv K' n e^{-ny}.$$

Likewise,

$$\frac{\partial u}{\partial y} : \quad \frac{\partial f_n}{\partial y}(x,y) = b_n \sin nx \frac{\cosh n(1-y)}{\sinh n}(-n).$$

However, working as we did for the bound (4.4.11), we easily find that

$$\left| \frac{\cosh n(1-y)}{\sinh n} \right| \le e^{-ny} \frac{2}{1-e^{-2}} \equiv H e^{-ny}$$

and therefore

$$\left|\frac{\partial f_n}{\partial y}(x,y)\right| \le |b_n| nHe^{-ny} \le \beta nHe^{-ny} \equiv H' ne^{-ny}.$$

Similarly one checks that the second-order derivatives are all dominated by $\sum_{n=1}^{\infty} n^2 e^{-ny}$.

3. Uniform bounds in the whole domain is sometimes too much

Evidently, there are no chances of dominating a series of the form $\sum_{n=1}^{\infty} n^2 e^{-ny}$ with a convergent numerical series in the whole rectangle $R =]0,\pi[\times]0,+\infty[$. However, this is not strictly necessary to get the desired regularity of u. To understand this point, it is enough to consider functions of one variable, taking for instance the series

$$\sum_{n=1}^{\infty} e^{-nx}.$$

This series converges pointwisely in $I =]0,+\infty[$ (thus defining a sum function f on the whole of I) and **converges uniformly for** $x \ge x_0$, **whatever is** $x_0 > 0$; indeed,

$$\sum_{n=1}^{\infty} e^{-nx} \le \sum_{n=1}^{\infty} e^{-nx_0} \equiv \sum_{n=1}^{\infty} M_n < +\infty \quad (x \in I_{x_0} \equiv [x_0,+\infty[).$$

This is enough to conclude that f is continuous in the whole of I: for any given $\hat{x} \in I$, choose an x_0 with $0 < x_0 < \hat{x}$; by the above, f is continuous in $I_{x_0} = [x_0,+\infty[$ and therefore in $\hat{x}$. The series of the derivatives is $\sum_{n=1}^{\infty}(-n)e^{-nx}$, and as

$$\sum_{n=1}^{\infty} |-ne^{-nx}| \le \sum_{n=1}^{\infty} ne^{-nx_0} \equiv \sum_{n=1}^{\infty} K_n < +\infty \quad (x \in I_{x_0} = [x_0,+\infty[),$$

it follows, reasoning as before, that $f \in C^1(I)$. Iterating this procedure, one concludes in fact that $f \in C^{\infty}(I)$.

Note that a similar reasoning is required, for instance, to ensure that the sum f of a power series $\sum_{n=0}^{\infty} a_n x^n$ is of class C^{∞} in its interval of convergence (Theorem 2.7.2).

4.5 The Cauchy–Dirichlet problem for the heat equation

We first consider the heat equation in one space dimension. That is, we take $\Omega =]a,b[\subset \mathbb{R}$, and putting for convenience $]a,b[=]0,\pi[$ we consider specifically the **Cauchy–Dirichlet** problem in $R = \Omega \times]0,+\infty[$ (see Figure 4.7)

$$(\mathbf{CD}) \begin{cases} \frac{\partial u}{\partial t} - \frac{\partial^2 u}{\partial x^2} = 0 & 0 < x < \pi, t > 0 \quad (\mathbf{E}) \\ u(x,0) = u_0(x) & 0 \le x \le \pi \quad (\mathbf{IC}) \\ u(0,t) = u(\pi,t) = 0 & t \ge 0, \quad (\mathbf{BC}) \end{cases} \tag{4.5.1}$$

where $u_0 \in C_0[0,\pi]$.

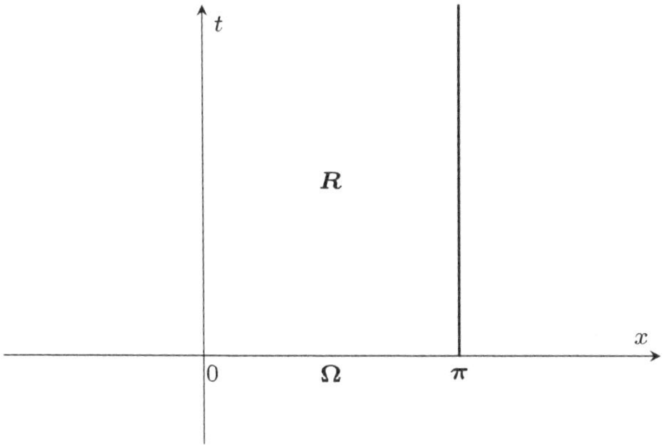

Figure 4.7: $R = \Omega \times]0, +\infty[$.

In the above system (**CD**), $u(x,t)$ has the physical meaning of the temperature at the point x and time t in a metallic wire occupying the segment $[0,\pi]$ of the x-axis, u_0 representing the initial distribution of the temperature in the wire. The endpoints of the wire are held at temperature 0 for all time t, and the absence of a term $f \neq 0$ on the right-hand side of the differential equation means that no external heat source is being provided to our one-dimensional body.

As we did for the problem (**D**), it can be shown (see, e. g., [19]) that if $u_0 = 0$ – that is, for the homogeneous problem associated with (**CD**) – the only solution is $u = 0$, so that (**CD**) has **at most one solution**. We are going to actually find it by the same techniques of separation of variables and Fourier series.

A. Separation of variables+boundary condition+series solution

Seeking elementary solutions of the differential equation of the form $u(x,t) = X(x)T(t)$ that also satisfy the boundary condition (**BC**) and then making an "infinite superposition" of these, one finds as candidate solution

$$u(x,t) = \sum_{n=1}^{\infty} b_n \sin nx e^{-n^2 t} \equiv \sum_{n=1}^{\infty} b_n v_n(x,t), \tag{4.5.2}$$

with $(x,t) \in \overline{R} = [0,\pi] \times [0,+\infty[$. See below for a full and more general discussion of this point.

B. Imposition of the initial condition

Imposing formally (**IC**) in (4.5.2), we find

$$u(x,0) = \sum_{n=1}^{\infty} b_n \sin nx = u_0(x) \quad (0 \le x \le \pi). \tag{4.5.3}$$

Choose the coefficients b_n

We have

$$b_n = \frac{2}{\pi}\int_0^{\pi} u_0(x)\sin nx\,dx \quad (n \in \mathbb{N}), \tag{4.5.4}$$

so that the series in (4.5.3) is the sine Fourier series of u_0.

Proposition 4.5.1. *Let $u_0 \in C_0^1[0,\pi]$ and let (b_n) be as in (4.5.4). Then the series in (4.5.3) converges pointwisely to u_0 in $[0,\pi]$; that is, (4.5.3) is satisfied. Moreover, the convergence of the series in $[0,\pi]$ is **uniform**. Finally, also the series in two variables*

$$\sum_{n=1}^{\infty} b_n \sin nx e^{-n^2 t}$$

converges uniformly in $\overline{R} = [0,\pi] \times [0,+\infty[$.

Proof. The proof is identical to that of Propositions 4.4.2 and 4.4.3 and Corollary 4.4.1; in fact, it is easier, because here we have immediately

$$|b_n \sin nx e^{-n^2 t}| \le |b_n| \quad \forall (x,t) \in [0,\pi] \times [0,+\infty[. \qquad \square$$

Theorem 4.5.1. *Let $u_0 \in C_0^1[0,\pi]$ and let u be the function defined in $\overline{R} = [0,\pi] \times [0,+\infty[$ putting*

$$u(x,t) = \sum_{n=1}^{\infty} b_n \sin nx e^{-n^2 t}, \quad b_n = \frac{2}{\pi}\int_0^{\pi} u_0(x)\sin nx\,dx. \tag{4.5.5}$$

Then u is (the) solution of (**CD**).

Proof. (i) u is continuous in $\overline{R}$ by Proposition 4.5.1 and by the properties of uniform convergence. Moreover, u satisfies by construction the initial condition (**IC**) and the boundary condition (**BC**).

(ii) In the open rectangle R, u is of class C^2 with respect to x and of class C^1 with respect to t and satisfies the differential equation (**E**); indeed,

$$u = \sum_{n=1}^{\infty} b_n v_n \quad \text{with } \frac{\partial v_n}{\partial t} - \frac{\partial^2 v_n}{\partial x^2} = 0 \quad \forall n \in \mathbb{N},$$

so that, by term-by-term differentiation of the series (4.5.5), we will also have $\frac{\partial u}{\partial t} - \frac{\partial^2 u}{\partial x^2} = 0$. The possibility of term-by-term differentiation is checked by the same means indicated in the proof of Theorem 4.4.1 and is based on the fact that the series for $\frac{\partial u}{\partial t}, \frac{\partial u}{\partial x}, \frac{\partial^2 u}{\partial x^2}$ are all dominated by

$$\sum_{n=1}^{\infty} n^2 e^{-n^2 t}$$

and therefore converge uniformly in $R_{t_0} \equiv [0,\pi] \times [t_0, +\infty[$, whatever is $t_0 > 0$. □

Exercise 4.5.1. Prove that

$$u(x,t) \to 0 \quad \text{as } t \to +\infty, \qquad \forall x \in [0,\pi]. \tag{4.5.6}$$

Extension to any number of space variables

The method of constructing a solution by separation of variables and Fourier series can be partly extended to any number of space variables. We see here the main points. Let Ω be a bounded, open, connected subset of $\mathbb{R}^n$ and consider the heat equation

$$\frac{\partial u}{\partial t} - \Delta_x u = f(\mathbf{x},t) \quad (\mathbf{x},t) \in \Omega \times]0,+\infty[, \tag{4.5.7}$$

where t is the time variable, while Δ_x stands for the Laplacian in the space variable $\mathbf{x} = (x_1, \dots, x_n)$:

$$\Delta_x = \sum_{i=1}^{n} \frac{\partial^2 u}{\partial x_i^2}.$$

We deal in particular with the **Cauchy–Dirichlet problem** for the (homogeneous) heat equation, which consists in finding a function $u = u(\mathbf{x},t) : \overline{\Omega} \times [0,+\infty[\to \mathbb{R}$ such that

$$\textbf{(CD)} \begin{cases} \frac{\partial u}{\partial t} - \Delta_x u = 0 & \mathbf{x} \in \Omega,\ t > 0 \quad \textbf{(E)} \\ u(\mathbf{x},0) = u_0(\mathbf{x}) & \mathbf{x} \in \Omega \quad \textbf{(IC)} \\ u(\mathbf{x},t) = 0 & \mathbf{x} \in \partial\Omega,\ t \geq 0, \quad \textbf{(BC)} \end{cases} \tag{4.5.8}$$

where u_0 (the *initial data*) is a given function in $C(\overline{\Omega})$ with $u_0 = 0$ on $\partial\Omega$.

Let us start with (**E**). From now on, we simply write Δ rather than Δ_x. Seeking solutions of the form $u(\mathbf{x},t) = X(\mathbf{x})T(t)$ and assuming that $u(\mathbf{x},t) \neq 0$ for $(\mathbf{x},t) \in \Omega \times]0,+\infty[$, one finds

$$\frac{\partial u}{\partial t} - \Delta u = 0 \Leftrightarrow XT' - (\Delta X)T = 0 \Leftrightarrow \frac{T'}{T} - \frac{\Delta X}{X} = 0$$
$$\Leftrightarrow \frac{T'}{T} = \frac{\Delta X}{X} = \text{const.} \equiv -\lambda.$$

Therefore,

$$\textbf{(E)} \Longleftrightarrow \begin{cases} \Delta X + \lambda X = 0 & \mathbf{x} \in \Omega \\ T' + \lambda T = 0 & t \in]0,+\infty[. \end{cases}$$

These are two **coupled** equations, for λ is the same in the two equations.

The second equation has for any λ the solutions

$$T(t) = Ce^{-\lambda t} \quad (C \in \mathbb{R}).$$

Consider now the initial and boundary conditions in (**CD**):

$$\begin{cases} u(\mathbf{x}, 0) = X(\mathbf{x})T(0) = u_0(\mathbf{x}) & \mathbf{x} \in \Omega & \textbf{(IC)} \\ u(\mathbf{x}, t) = X(\mathbf{x})T(t) = 0 & \mathbf{x} \in \partial\Omega,\ t \geq 0. & \textbf{(BC)} \end{cases} \tag{4.5.9}$$

Equation (**BC**) is satisfied if $X(\mathbf{x}) = 0 \quad \forall \mathbf{x} \in \partial\Omega$.

Summing up, we see that (**E**) + (**BC**) are satisfied by $u(\mathbf{x}, t) = X(\mathbf{x})T(t)$ if $T(t) = Ce^{-\lambda t}$ and X is a solution of the problem

$$(\mathbf{EV}) \begin{cases} \Delta X + \lambda X = 0 & \text{in } \Omega \\ X = 0 & \text{on } \partial\Omega. \end{cases}$$

Question. Do there exist **non-zero** solutions of (**EV**)? More precisely, do there exist values of λ such that (**EV**) has solutions $X \neq 0$?

In this case, they are called respectively **eigenvalues** and **eigenfunctions** of the Laplace operator (the *Laplacian*) Δ in Ω with Dirichlet boundary conditions (more precisely, of $-\Delta$, for the equation in (**EV**) can be written in the form $-\Delta X = \lambda X$, which is more consistent with the usual way of writing eigenvalue problems).

We have answered this question in case $n = 1$, where it is possible to solve explicitly (**EV**); we have found indeed (Exercise 3.5.1) that the problem

$$\begin{cases} -X'' = \lambda X & \text{in }]a, b[\\ X(a) = X(b) = 0 \end{cases}$$

has the following eigenvalues and eigenfunctions:

$$\lambda_k = \left(\frac{k\pi}{b-a}\right)^2, \quad X_k(x) = C \sin \omega_k (x - a) \quad (k \in \mathbb{N}),$$

with $\omega_k = \sqrt{\lambda_k}$.

A classical and fundamental result in both mathematical analysis and mathematical physics (see, e. g., [14] or [13]) ensures that, essentially, this remains true in any space dimension.

Theorem 4.5.2. *There exists an infinite sequence of positive eigenvalues λ_k of* (**EV**). *They form a strictly increasing sequence going to $+\infty$:*

$$0 < \lambda_1 < \lambda_2 < \cdots < \lambda_k < \lambda_{k+1} \ldots, \quad \lambda_k \to \infty \ (k \to \infty).$$

Let X_k denote a norm-one solution corresponding to λ_k. Then the sequence (X_k) of the normalized eigenfunctions of the (Dirichlet) Laplacian in Ω forms an orthonormal sequence in $C(\overline{\Omega})$, this space being equipped with the inner product

$$\langle f, g\rangle = \int_{\overline{\Omega}} f(\boldsymbol{x})g(\boldsymbol{x})\,d\boldsymbol{x}. \tag{4.5.10}$$

Moreover, the sequence (X_k) is total *in this space, so that every function $u \in C(\overline{\Omega})$ can be expanded in series (in the sense of the convergence in the quadratic mean on $\overline{\Omega}$) of the eigenfunctions X_k.*

Conclusion (about (CD))

For every $k \in \mathbb{N}$, the function

$$u_k(\boldsymbol{x}, t) = X_k(\boldsymbol{x})e^{-\lambda_k t}$$

is a solution of (**E**) + (**BC**). Therefore (see “General Remark” **3.**), we look for a solution of the form

$$u(\boldsymbol{x}, t) = \sum_{k=1}^{\infty} c_k u_k(\boldsymbol{x}, t) = \sum_{k=1}^{\infty} c_k X_k(\boldsymbol{x})e^{-\lambda_k t}, \tag{4.5.11}$$

where the constants $c_k (k \in \mathbb{N})$ have to be chosen in such a way that:
- the series (4.5.11) is (uniformly) convergent in $\overline{\Omega} \times [0, +\infty[$;
- the condition (**IC**) is satisfied.

Imposing formally (**IC**), we find

$$u(\boldsymbol{x}, 0) = \sum_{k=1}^{\infty} c_k X_k(\boldsymbol{x}) = u_0(\boldsymbol{x}), \quad \boldsymbol{x} \in \Omega. \tag{4.5.12}$$

The possibility of satisfying this condition and determining the constants c_k amounts to the possibility of **expanding u_0 in series of the eigenfunctions X_k**. This is granted by our assumption that $u_0 \in C(\overline{\Omega})$ and by the totality of the (X_k) stated in Theorem 4.5.2. Note that – as should be clear from the statement of this theorem – the **eigenfunction expansion** stated above refers to the convergence of the series in the **quadratic mean** induced by the inner product (4.5.10).

Remark 4.5.1. The proof of Theorem 4.5.2 requires advanced results from functional analysis [14] involving the spectral properties of compact symmetric operators in Hilbert spaces, as sketched in Theorem 3.5.2 after the discussion of Sturm–Liouville eigenvalue problems. Here, however, handling functions of several variables requires a much heavier technical apparatus, starting with the construction and properties of the appropriate Sobolev spaces.

4.6 Additions and exercises

A1. The wave equation

As to the wave equation, one might in general consider the following IVP/BVP: given an open set $\Omega \subset \mathbb{R}^n$ with boundary $\partial\Omega$, find a function

$$u = u(\mathbf{x}, t) : \overline{\Omega} \times [0, +\infty[\to \mathbb{R}$$

satisfying the following **Cauchy–Dirichlet problem**:

$$\begin{cases} \frac{\partial^2 u}{\partial t^2} - \Delta_x u = f(\mathbf{x}, t) & \text{in } \Omega \times]0, +\infty[\\ u(\mathbf{x}, 0) = u_0(\mathbf{x}) & \mathbf{x} \in \Omega \\ u_t(\mathbf{x}, 0) = v_0(\mathbf{x}) & \mathbf{x} \in \Omega \\ u(\mathbf{x}, t) = g(\mathbf{x}, t) & \mathbf{x} \in \partial\Omega, t \geq 0. \end{cases} \tag{4.6.1}$$

In (4.6.1), f, g, u_0, v_0 are given functions, with u_0 and v_0 representing the **initial conditions** (Cauchy conditions) imposed on the unknown function u, while the last condition is a **Dirichlet boundary condition** imposed on u; it could be replaced for instance by the **Neumann** boundary condition, where the exterior normal derivative of u on $\partial\Omega$ is assigned instead of u.

Here we will only say something on the Cauchy–Dirichlet problem for the **homogeneous** wave equation and with zero Dirichlet boundary condition, that is, on the following simplified form of (4.6.1):

$$(\mathbf{CD}) \begin{cases} \frac{\partial^2 u}{\partial t^2} - \Delta_x u = 0 & x \in \Omega,\ t > 0 & (\mathbf{E}) \\ u(x, 0) = u_0(x) & x \in \Omega & (\mathbf{IC1}) \\ \frac{\partial u}{\partial t}(x, 0) = v_0(x) & x \in \Omega & (\mathbf{IC2}) \\ u(x, t) = 0 & x \in \partial\Omega,\ t \geq 0, & (\mathbf{BC}) \end{cases} \tag{4.6.2}$$

where the initial data u_0 and v_0 are given function in $C(\overline{\Omega})$, with $u_0 = 0$ on $\partial\Omega$.

Physical interpretation

As already mentioned in Section 4.1 of this Chapter, for $n = 2$ equation **(E)** models the motion of a vibrating **membrane** on which no external force is exerted, and **(BC)** means that the membrane is held fixed on its boundary for all time. Likewise if $n = 1$, in which case **(E)** reduces to

$$\frac{\partial^2 u}{\partial t^2} - \frac{\partial^2 u}{\partial x^2} = 0 \quad x \in \Omega =]a, b[,\ t > 0. \tag{4.6.3}$$

This represents the vibrations of an elastic **string** without external force; $u(x, t)$ represents the vertical displacement ("displacement along the y-axis") at time t of that point of the string that at rest has coordinate $x \in \Omega$; thus, in the plane (x, y), at time t the configuration of the string can be described by the arrow

$$x \equiv (x, 0) \to (x, u(x, t)), \quad a \le x \le b,$$

that is, by the *graph* of the function $u(., t) : x \to u(x, t) : [a, b] \to \mathbb{R}$. In the full problem **(CD)**, $u_0 = u_0(x)$ and $v_0 = v_0(x)$ represent respectively the initial position and velocity of the string, while – if $\Omega =]a, b[$ is bounded – **(BC)** writes $u(a, t) = u(b, t) = 0$ for all t, which means that the endpoints of the string are fixed on the x-axis ($y = 0$) for all time.

D'Alembert solution of the wave equation

Remaining for a while in the case $n = 1$, it is interesting and historically important to look first at **(CD)** when $\Omega = \mathbb{R}$, in which case the boundary condition vanishes. Recalling that the solutions of (4.6.3) are of the form

$$u(x, t) = f(x + t) + g(x - t), \tag{4.6.4}$$

with $f, g : \mathbb{R} \to \mathbb{R}$ arbitrary (it is often said that every solution u is the sum of a **progressing** wave and of a **regressing** wave), D'Alembert found the solution

$$u(x, t) = \frac{1}{2}[u_0(x + t) + u_0(x - t)] + \int_{x-t}^{x+t} v_0(s)\, ds \tag{4.6.5}$$

of **(CD)**. It is clear that (4.6.5) is of the form (4.6.4), and it is easily checked that the initial conditions **(IC1)** and **(IC2)** are also satisfied.

If $\Omega =]a, b[$ is bounded, the D'Alembert solution (4.6.5) can still be used to construct a solution of **(CD)**; this requires careful and rather lengthy manipulations that are shown for instance in Weinberger [2] and in Pagani–Salsa [19].

Solution of (CD) by eigenfunction expansion

Let us now briefly discuss the case $n \ge 1$ proceeding by separation of variables and series expansion as we did with the heat equation. Assuming that $u(\mathbf{x}, t) = X(\mathbf{x})T(t)$ and writing Δ rather than Δ_x, we find that (**E**) splits into the system

$$\begin{cases} \Delta X + \lambda X = 0 & \mathbf{x} \in \Omega \\ T'' + \lambda T = 0 & t \in]0, +\infty[. \end{cases}$$

Consider first in (4.6.2) the boundary condition (**BC**); this is satisfied by $u(\mathbf{x}, t) = X(\mathbf{x})T(t)$ if $X(\mathbf{x}) = 0\ \forall \mathbf{x} \in \partial\Omega$, irrespective of T. Therefore, as to X, again we have to deal with the eigenvalue problem

$$(\mathbf{EV})\begin{cases} \Delta X + \lambda X = 0 & \text{in } \Omega \\ X = 0 & \text{on } \partial\Omega, \end{cases}$$

and assuming the truth of the spectral theorem for the Dirichlet Laplacian, Theorem 4.5.2, we find the existence of infinitely many distinct eigenvalues $\lambda_k > 0\,(k \in \mathbb{N})$ with corresponding eigenfunctions X_k. The eigenvalues λ_k are the only values of the parameter λ that matter in our construction, for otherwise the unique solution of (**EV**) is $X = 0$, making $u(\mathbf{x}, t) = X(\mathbf{x})T(t) = 0$, which of course does not solve our problem (4.6.2) except in the trivial case in which $u_0 = v_0 = 0$. The differential equation that we must consider for T is therefore

$$T'' + \lambda_k T = 0, \quad t > 0,$$

whose general solution is, for each fixed $k \in \mathbb{N}$,

$$T(t) = C\cos\mu_k t + D\sin\mu_k t, \quad \mu_k = \sqrt{\lambda_k}, \quad C, D \in \mathbb{R}.$$

Summing up, the **elementary solutions** of (**E**) + (**BC**) in our problem (4.6.2) are given by

$$u_k(\mathbf{x}, t) = [C_k \cos\mu_k t + D_k \sin\mu_k t]X_k(\mathbf{x}), \quad k \in \mathbb{N}, \tag{4.6.6}$$

and we search as usual the solution in the form

$$u(\mathbf{x}, t) = \sum_{k=1}^{\infty} u_k(\mathbf{x}, t). \tag{4.6.7}$$

Impose (at least formally) the initial conditions (**IC1**) and (**IC2**):

$$u(\mathbf{x}, 0) = \sum_{k=1}^{\infty} C_k X_k(\mathbf{x}) = u_0(\mathbf{x}), \tag{4.6.8}$$

$$\frac{\partial u}{\partial t}(\mathbf{x}, 0) = \sum_{k=1}^{\infty} D_k \mu_k X_k(\mathbf{x}) = v_0(\mathbf{x}). \tag{4.6.9}$$

These two conditions demand that u_0 and v_0 be expanded in series of the eigenfunctions X_k of the Laplacian, and this is ensured – at least as to the convergence in the quadratic mean – by our assumptions that u_0 and v_0 are in $C(\overline{\Omega})$ and by the completeness of the (X_k). Then the above equations (4.6.8) and (4.6.9) determine uniquely the coefficients C_k and D_k, respectively, and therefore the solution u via (4.6.6) and (4.6.7). The quality of the convergence of the series in (4.6.7) (ensuring that its sum is actually a solution of (**CD**)) will – roughly speaking – depend on the higher regularity of u_0 and v_0.

Exercises

E1. Solutions of some of the exercises given in the text

Section 4.1

Exercise 1.1 (Fundamental solutions of the Laplace equation)

Case $N = 2$

Consider

$$u(x,y) = \ln\sqrt{x^2+y^2}, \quad (x,y) \neq (0,0). \tag{4.6.10}$$

For $(x,y) \neq (0,0)$, we have

$$\frac{\partial}{\partial x}\ln\sqrt{x^2+y^2} = \frac{1}{\sqrt{x^2+y^2}}\frac{\partial}{\partial x}\sqrt{x^2+y^2} = \frac{x}{x^2+y^2},$$

whence

$$\frac{\partial^2}{\partial x^2}\ln\sqrt{x^2+y^2} = \frac{\partial}{\partial x}\left[\frac{x}{x^2+y^2}\right] = \frac{(x^2+y^2)-x.2x}{(x^2+y^2)^2} = \frac{y^2-x^2}{(x^2+y^2)^2}.$$

Similarly,

$$\frac{\partial^2}{\partial y^2}\ln\sqrt{x^2+y^2} = \frac{x^2-y^2}{(x^2+y^2)^2},$$

so that

$$\left(\frac{\partial^2}{\partial x^2}+\frac{\partial^2}{\partial y^2}\right)\ln\sqrt{x^2+y^2} = 0.$$

Case $N = 3$

Consider

$$u(x,y,z) = \frac{1}{\sqrt{x^2+y^2+z^2}} = \frac{1}{\|\mathbf{x}\|}, \quad \mathbf{x} \neq \mathbf{0}. \tag{4.6.11}$$

Put $r = r(\mathbf{x}) = \|\mathbf{x}\|$, so that $u(\mathbf{x}) = \frac{1}{r} = \frac{1}{r(\mathbf{x})}$. Then by the chain rule,

$$\frac{\partial u}{\partial x} = \frac{\partial u}{\partial r}\frac{\partial r}{\partial x} = -\frac{1}{r^2}\frac{\partial r}{\partial x} = -\frac{x}{r^3}$$

because

$$\frac{\partial r}{\partial x} = \frac{\partial}{\partial x}\sqrt{x^2+y^2+z^2} = \frac{x}{r}.$$

Moreover,

$$\frac{\partial^2 u}{\partial x^2} = \frac{\partial}{\partial x}\left(\frac{\partial u}{\partial x}\right) = -\frac{\partial}{\partial x}\left(\frac{x}{r^3}\right) = -\left[\frac{1}{r^3} + x\frac{\partial}{\partial x}\left(\frac{1}{r^3}\right)\right]$$
$$= -\left[\frac{1}{r^3} - \frac{3x}{r^4}\frac{\partial r}{\partial x}\right] = -\frac{1}{r^3} + \frac{3x^2}{r^5},$$

similar expressions holding for the remaining partial derivatives. Summing up, for $(x, y, z) \neq (0, 0, 0)$ we have

$$\frac{\partial^2 u}{\partial x^2} + \frac{\partial^2 u}{\partial y^2} + \frac{\partial^2 u}{\partial z^2} = -\frac{3}{r^3} + \frac{3}{r^5}(x^2 + y^2 + z^2) = -\frac{3}{r^3} + \frac{3}{r^3} = 0.$$

Similar computations show that in fact, for any $N \geq 3$, the function

$$u(\mathbf{x}) = \frac{1}{\|\mathbf{x}\|^{N-2}} = \frac{1}{r^{N-2}}$$

solves the Laplace equation in $\mathbb{R}^N \setminus \{0\}$.

Exercise 1.2

Iterating the computations based on the chain rule and displayed after equation (4.1.18), we find that

$$\frac{\partial^2 u}{\partial t^2} - v^2\frac{\partial^2 u}{\partial x^2} = -4v^2\frac{\partial^2 \omega}{\partial p \partial q},$$

showing that in the new variables p, q the one-dimensional wave equation (4.1.17) becomes

$$\frac{\partial^2 \omega}{\partial p \partial q} = 0,$$

whence $\omega(p, q) = f(p) + g(q)$ by (4.1.16), and therefore

$$u(x, t) = \omega(x + tv, x - tv) = f(x + tv) + g(x - tv).$$

Vice versa, if $u(x, t) = f(x + tv) + g(x - tv)$, where $f, g : \mathbb{R} \to \mathbb{R}$ are twice differentiable arbitrary functions, then

$$\frac{\partial^2 u}{\partial x^2} = f''(x + tv) + g''(x + tv)$$

and

$$\frac{\partial^2 u}{\partial t^2} = f''(x + tv)v^2 + g''(x + tv)v^2,$$

whence

$$\frac{\partial^2 u}{\partial t^2} - v^2 \frac{\partial^2 u}{\partial x^2} = 0 \quad \forall (t, x) \in \mathbb{R}^2.$$

Exercise 1.3

Let $v \in C^1(\Omega)$, $\mathbf{z} \in C^1(\Omega, \mathbb{R}^n)$. Writing $\mathbf{z} = (z_1, \ldots, z_n)$, we have

$$\operatorname{div}(v\mathbf{z}) = \sum_{i=1}^{n} \frac{\partial}{\partial x_i}(vz_i) = \sum_{i=1}^{n} \left(\frac{\partial v}{\partial x_i} z_i + v \frac{\partial z_i}{\partial x_i} \right) = \nabla v \cdot \mathbf{z} + v(\operatorname{div} \mathbf{z}).$$

Section 4.3

Exercise 3.1

To solve this exercise, follow the pattern shown in Exercise 3.2 below, save that the two functions indicated in the text of the exercise are *not* piecewise of class C^1, and thus Proposition 4.3.1 cannot be used. In particular, the presence of discontinuities rules out the possibility that the convergence of their Fourier series be uniform.

Exercise 3.2

(a) First compute the Fourier coefficients of the 2π-periodic extension f of the function $x \to x^2$, $|x| \le \pi$. We have $b_k = 0$ for every $k \in \mathbb{N}$, while

$$a_0 = \frac{1}{\pi} \int_0^{\pi} x^2 \, dx = \frac{\pi^2}{3}$$

and

$$a_k = \frac{2}{\pi} \int_0^{\pi} x^2 \cos kx \, dx = \frac{4}{k^2}(-1)^k \quad (k \in \mathbb{N}),$$

as follows for instance by repeated integration by parts:

$$\begin{aligned} \int_0^{\pi} x^2 \cos kx \, dx &= \frac{1}{k}[x^2 \sin x]_0^{\pi} + \frac{2}{k^2}[x \cos kx]_0^{\pi} - \frac{2}{k^2}[\sin kx]_0^{\pi} \\ &= \frac{2}{k^2} \pi \cos k\pi = \frac{2}{k^2} \pi (-1)^k. \end{aligned}$$

(b) In order to prove the pointwise convergence of the Fourier series of f to f itself, and thus in particular the validity of the formula

$$x^2 = \frac{\pi^2}{3} + 4 \sum_{k=1}^{\infty} \frac{(-1)^k}{k^2} \cos kx, \quad x \in [-\pi, \pi], \tag{4.6.12}$$

we should check that the hypotheses of Theorem 4.3.1 are satisfied. Now, our f is continuous on all of $\mathbb{R}$ and differentiable at each point of $\mathbb{R}$ except at the points

$$\pi \pm 2k\pi \quad (k \in \mathbb{N}).$$

However, at these points, the left- and right-derivatives of f exist; this can be checked directly by the definitions or via Remark 4.3.3, as we have for instance

$$\lim_{x\to\pi^-} f'(x) = \lim_{x\to\pi^-} 2x = 2\pi, \qquad \lim_{x\to\pi^+} f'(x) = \lim_{x\to\pi^+} 2(x-2\pi) = -2\pi.$$

We have thus verified that f is **piecewise of class C^1 on $\mathbb{R}$** (meaning that it is so on each interval $[a,b] \subset \mathbb{R}$), and the pointwise convergence to f of the series at the right-hand side of (4.6.12) is thus proved via Proposition 4.3.1.

In fact, the very same properties of f imply, via Theorem 4.3.2, that the convergence is **uniform**.

Section 4.5

Exercise 5.1 (Exponential decay of the temperature)

Recall that

$$u(x,t) = \sum_{n=1}^{\infty} b_n \sin nx e^{-n^2 t}, \quad b_n = \frac{2}{\pi}\int_0^{\pi} u_0(x) \sin nx \, dx. \tag{4.6.13}$$

Thus, we have

$$|u(x,t)| \le C \sum_{n=1}^{\infty} e^{-n^2 t}$$

for some $C > 0$ and for all $(x,t) \in [0,\pi] \times [0,+\infty[$. Now fix a $t_0 > 0$ and write, for $t > t_0$,

$$e^{-n^2 t} = e^{-n^2(t-t_0)} e^{-n^2 t_0} \le e^{-(t-t_0)} e^{-n^2 t_0},$$

whence

$$|u(x,t)| \le C e^{-(t-t_0)} \sum_{n=1}^{\infty} e^{-n^2 t_0} \equiv K e^{-(t-t_0)}$$

for $t > t_0$ and for all $x \in [0,\pi]$. This shows that the asymptotic behavior (4.5.6) of the temperature holds with an **exponential decay**, and **uniformly** with respect to $x \in [0,\pi]$.

Bibliography

[1] Walter, W. *Ordinary Differential Equations*, Springer, 1998.
[2] Weinberger, H. F. *A first course in Partial Differential Equations*, Blaisdell, 1965.
[3] Ahmad, S; Ambrosetti, A. *Differential Equations*, De Gruyter, 2019.
[4] Apostol, T. M. *Calculus* (2 Volumes), Wiley, 1969.
[5] Hale, J. K. *Ordinary Differential Equations*, Wiley, 1969.
[6] Rudin, W. *Real and Complex Analysis*, McGraw-Hill, 1966.
[7] Dieudonné, J. *Foundations of Modern Analysis*, Academic Press, 1969.
[8] Taylor, A. E. *Introduction to Functional Analysis*, Wiley, 1958.
[9] Kolmogorov, A.; Fomin, S. *Eléments de la théorie des fonctions et de l'analyse fonctionnelle*, MIR, 1977.
[10] Rudin, W.: *Principles of Mathematical Analysis*, McGraw-Hill, 1976.
[11] Edmunds, D. E.; Evans, W. D. *Spectral Theory and Differential Operators*, Oxford University Press, 2018.
[12] Coddington, E. A.; Levinson, N. *Theory of Ordinary Differential Equations*, McGraw-Hill, 1955.
[13] Courant, R.; Hilbert, D. *Methods of Mathematical Physics* (2 Volumes), Interscience, 1962.
[14] Brezis, H. *Functional Analysis, Sobolev Spaces and Partial Differential Equations*, Springer, 2011.
[15] Lions, J. L. *Quelques méthodes de résolution des problèmes aux limites non linéaires*, Dunod-Gauthier Villars, 1969.
[16] Trèves, F. *Basic Linear Partial Differential Equations*, Academic Press, 1975.
[17] Gilbarg, D.; Trudinger, N. S. *Elliptic Partial Differential Equations of Second Order*, Springer, 1977.
[18] Fleming, W. *Functions of Several Variables*, Springer, 1977.
[19] Pagani, C. D.; Salsa, S. *Analisi Matematica 2*, Zanichelli, 2016.

https://doi.org/10.1515/9783111302522-005

Index

https://doi.org/10.1515/9783111302522-006

www.ingramcontent.com/pod-product-compliance
Lightning Source LLC
Chambersburg PA
CBHW061416210326
41598CB00035B/6235
9783111294858